# HARVEY'S ELECTRICAL CODE FIELD GUIDE, 2nd Edition

## Unit #1

## Formulas, Rules & Laws - AC & DC

## Unit #2

## Service Entrance, Feeders & Related Devices



## Unit #3

## Branch Circuits

## Unit #4

## Electric Motors & Motor Controls



## Unit #5

## Transformers

# Unit #6

## Electrical Terms, Color Codes & Symbols

## UNIT 1

## FORMULAS, RULES, AND LAWS - AC & DC

This "formula wheel" summarizes the formulas you will need to use to find voltage, amperage, resistance and power in an electrical circuit. The formulas are grouped in the quarter circles and apply to the four electrical values listed in the smaller inner circle. To use the wheel as a handy reference, find the unknown value in one of the four quadrants in the wheel's center circle. In the outer part of the same quadrant, find the letters of your two known circuit values. The wheel tells you what formula to use to solve for the unknown.

Figure 1.1

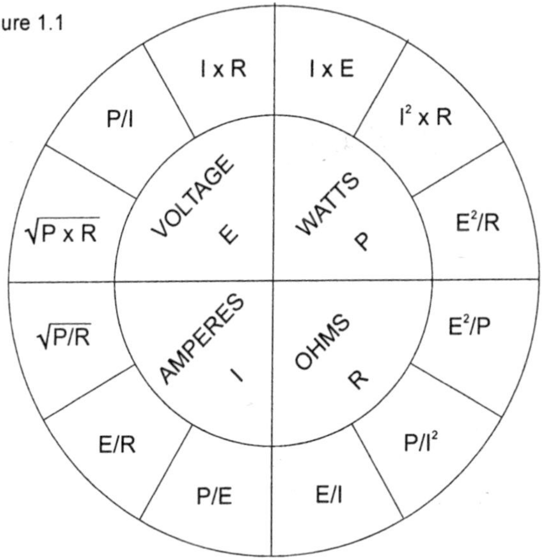


**Electrical Formula Wheel**

Formulas can also be summarized in table form, thus:

| FORMULA | SOURCE |
|---|---|
| 1. $E = IR$ | Ohm's Law |
| 2. $E = \frac{P}{I}$ | Power Law |
| 3. $E = \sqrt{PR}$ | Transposes No. 12 plus square root |
| 4. $I = \frac{E}{R}$ | Ohm's Law |
| 5. $I = \frac{P}{E}$ | Power Law |
| 6. $I = \sqrt{\frac{P}{R}}$ | Transposition of No. 9 plus square root |
| 7. $R = \frac{E}{I}$ | Ohm's Law |
| 8. $R = \frac{E^2}{P}$ | Transposition of Formula 12 |
| 9. $R = \frac{P}{I^2}$ | Transposition of Formula 11 |
| 10. $P = IE$ | Power Law |
| 11. $P = I^2R$ | Substituting IR, Formula 1 for E |
| 12. $P = \frac{E^2}{R}$ | Substituting $\frac{E}{R}$ from Formula 4 for I |

## OHM'S LAW

The basic relationship of current (amperage), electromotive force (voltage), and resistance in an electric circuit is expressed in Ohm's Law: The rate of flow of the current is equal to the electromotive force divided by the resistance. The formulas are:

$$E = IR \quad or \quad I = \frac{E}{R} \quad or \quad R = \frac{E}{I}$$

Where:

$$I = Current = Amperes\ (A)$$
$$E = Voltage = Volts\ (V)$$
$$R = Resistance = Ohms$$
$$P = Power = Watts\ (W)\ or\ Volt\text{-}Ampere\ (VA)$$

## ELECTRIC CIRCUITS

There are three types of electric circuits. Each has a set of rules that govern the action of current, voltage and resistance. The three types are:

* Series Circuits - Circuits having a single pathway.
* Parallel Circuits - Circuits having more than one path.
* Combination Circuits - Circuits that are made up of both series and parallel portions.

### Series Circuit Rules:

1. The total current in a series circuit equals the current in any part of the circuit:

$$I_{total} = I_1 = I_2 = I_3,\ etc.$$

2. Total voltage in a series circuit equals the sum of the voltages across all the parts of the circuit.

$$E_{total} = E_1 + E_2 + E_3, \text{ etc.}$$

3. Total resistance of the series circuit is equal to the sum of the resistances of all the parts of the circuit.

$$R_{total} = R_1 + R_2 + R_3, \text{ etc.}$$

**Example:**

Using the given values in the illustration following, find the missing currents, voltages, and resistances in a series circuit.

Figure 1.2

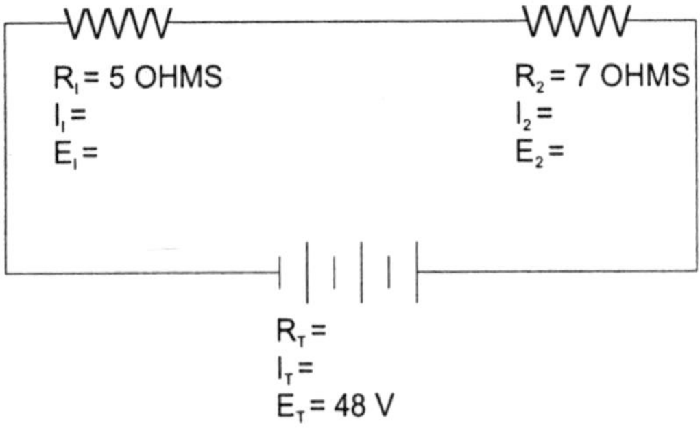


**Solution:**

$$R_T = R_1 + R_2 = 5 + 7 = 12 \, ohms$$

$$I_T = \frac{E_T}{R_T} = \frac{48}{12} = 4 \text{ amperes } \underline{\text{and}} = I_1 = I_2$$

$E_1 = I_1 \times R_1 = 4 \times 5 = 20$ volts

$E_2 = I_2 \times R_2 = 4 \times 7 = 28$ volts

## Parallel Circuit Rules:

1. Total current in a parallel circuit is equal to the sum of the currents in all the branches of the circuit.

$$I_{total} = I_1 + I_2 + I_3, \text{ etc.}$$

2. Total voltage across any branch of a parallel circuit is equal to the voltage across any other branch and is also equal to the total voltage.

$$E_{total} = E_1 = E_2 = E_3, \text{ etc.}$$

3. Total resistance of a parallel circuit is equal to the total voltage divided by the total current.

$$R_T = \frac{E_T}{I_T}$$

Also, total resistance of a parallel circuit may be determined by using the following formula:

$$\frac{1}{R_T} = \frac{1}{R_1} + \frac{1}{R_2} + \frac{1}{R_3}, \text{ etc.}$$

*Where R1 , R2 , R3, etc. are branch resistances.*

## Example:

Given the information shown in the following drawing, solve the parallel circuit for the unknowns.

Figure 1.3

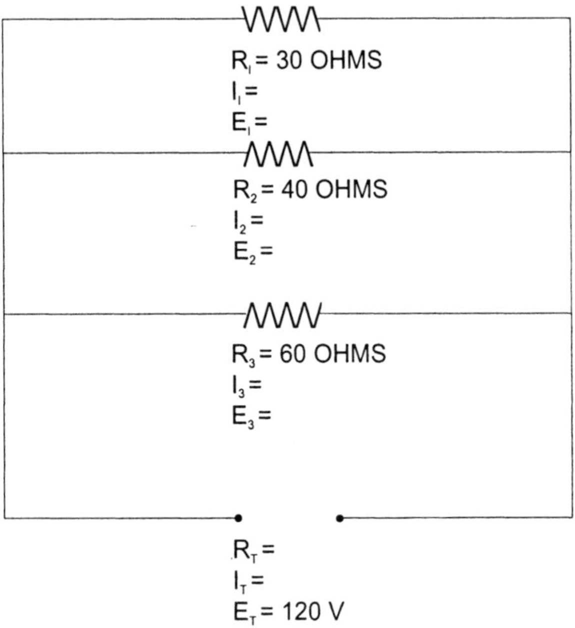


**Solution:**

$$\frac{1}{R_T} = \frac{1}{R_1} + \frac{1}{R_2} + \frac{1}{R_3} = \frac{1}{30} + \frac{1}{40} + \frac{1}{60}$$

$$= \frac{4 + 3 + 2^{*}}{120}$$

*Use the lowest common denominator

Thus: $\frac{1}{R_T} = \frac{9}{120}$ **OR**:

$$R_T = \frac{120}{9} = 13.33 \text{ ohms}$$

$$E_T = E_1 = E_2 = E_3$$

$$120 \text{ V} = E_1 = E_2 = E_3$$

$$I_T = \frac{E_T}{R_T} = \frac{120V}{13.33} = 9 \text{ amperes}$$

## Combination Circuit Rules:

Since this type of circuit is partly series and partly parallel, the circuit may be analyzed by applying the rules regarding series circuits to the series portion and those regarding parallel circuits to the parallel portion.

Also, Ohm's Law applies throughout the entire circuit.

## Example:

Given the information shown in the series-parallel circuit drawing following, solve for the unknowns in the diagram on the following page.

Figure 1.4

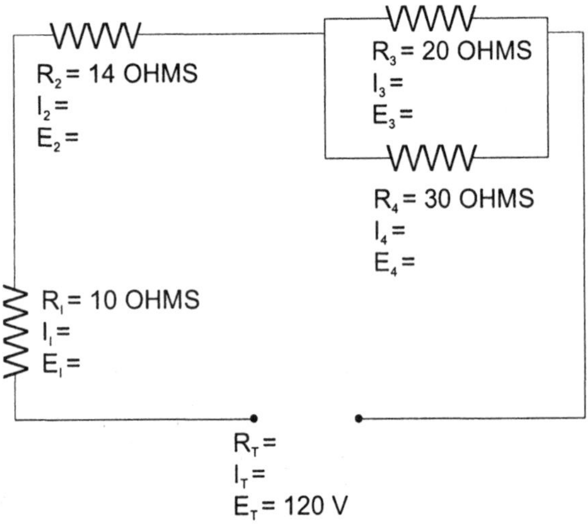


**Solution:**

Since $R_1$ and $R_2$ are in series, add them:

$R_1 + R_2 = 10 + 14 = 24$ ohms

Since $R_3$ and $R_4$ are in parallel: $\frac{1}{R_{3,4}} = \frac{1}{R_3} + \frac{1}{R_4} = \frac{1}{20} + \frac{1}{30}$

$= \frac{3}{60} + \frac{2}{60} = \frac{5}{60}$ (change to lowest common denominator)

$= \frac{60}{5} = 12$ (invert to divide). Thus, $R_{3,4} = 12$ ohms

Also, since $R_{3,4}$ are in series with $R_1$ and $R_2$, they can be added together:

$$R_T = R_1 + R_2 + R_{3,4} = 10 + 14 + 12 = 36 \text{ ohms}$$

$$I_T = \frac{E_T}{R_T} = \frac{120}{36} = 3.36 \text{ amperes} = I_1 = I_2 = I_{3,4}$$

Next: $E_1 = I_1 \times R_1 = 3.36 \times 10 = 33.6$ volts

And: $E_2 = I_2 \times R_2 = 3.36 \times 14 = 47$ volts

And: $E_{3,4} = I_{3,4} \times R_{3,4} = 3.36 \times 12 = 40.3$ volts

$$I_3 = \frac{E_3}{R_3} = \frac{40.3\,V}{20\,ohms} = 2.02 \text{ amperes}$$

$$I_4 = \frac{E_4}{R_4} = \frac{40.3\,V}{30\,ohms} = 1.34 \text{ amperes}$$

$$I_1 = \frac{E_1}{R_1} = \frac{120}{30} = 4 \text{ amperes}$$

$$I_2 = \frac{E_2}{R_2} = \frac{120}{40} = 3 \text{ amperes}$$

$$I_3 = \frac{E_3}{R_3} = \frac{120}{60} = 2 \text{ amperes}$$

$$I_T = I_1 + I_2 + I_3 = 4 + 3 + 2 = 9 \text{ amperes}$$

## ELECTRICAL POWER

Power of a circuit, measured in watts or more properly, volt-amperes, is the primary product of electrical energy. The basic formula is:

$$Power\ (P) = Voltage\ (E) \times Amperage\ (I)$$

One horsepower (hp) is equal to 746 watts or volt-amperes. One thousand watts (or volt-amperes) equals one kilowatt or one kilovolt-ampere.

The formula and values are usually stated as follows:

$P = E \times I$

1 hp = 746 W or VA

1000 W or VA = 1kW or 1kVA

### To determine horsepower (hp):

In a DC circuit:

NOTE: 1 horsepower is the amount of energy needed to lift 550 lb. one foot in one second.

$$Horsepower = volts \times amperes \times efficiency$$

### Example:

A 48 V motor uses a current of 2 amperes and has an efficiency of 95%. Find the horsepower of the motor.

### Solution:

$$HP = \frac{E \times I \times EFF}{746} = \frac{48 \times 2 \times 0.95}{746} = \frac{1}{8}\ \text{hp (approximately)}$$

In a single-phase circuit:

$$Horsepower = \frac{volts \times amperes \times efficiency \times power\ factor}{746}$$

**Example:**

A 120 V AC motor has an efficiency of 94% and a power factor of 85%. Find the horsepower if this motor draws 4.0 A.

**Solution:**

$$HP = \frac{E \times I \times EFF \times PF}{746}$$

$$= \frac{120 \times 4 \times 0.94 \times 0.85}{746}$$

$$= 0.5141 = \frac{1}{2} hp \ ^*$$

* *approximate value*

In a three-phase circuit:

$$HP = \frac{volts \times amperes \times EFF \times PF \times 1.73}{743}$$

**Example:**

A three-phase AC motor, 480 V, drawing 50 A, has an efficiency of 95% and a power factor of 88%. Find its approximate horsepower.

**Solution:**

$$HP = \frac{E \times I \times EFF \times PF \times 1.73}{746}$$

$$= \frac{480 \times 50 \times 0.95 \times 0.88 \times 1.73}{746}$$

$$= 46.5\, hp$$

NOTE: To determine kilowatts, use these same formulas but divide by 1000 rather than by 746.

**Example using the three-phase circuit:**

$$kW = \frac{E \times I \times EFF \times PF \times 1.73}{1000}$$
$$= \frac{480 \times 50 \times 0.95 \times 0.88 \times 1.73}{1000}$$
$$= 34.7 kW$$

**To determine kilovolt-amperes (kVA):**

In single-phase circuits:

$$Kilovolt\text{-}amperes = \frac{volts \times amperes}{1000}$$

**Example:**

A single-phase 240 V generating unit supplies 60 amperes at maximum output. Find its kVA.

**Solution:**

$$kVA = \frac{E \times I}{1000} = \frac{240 \times 60}{1000} = 14.4\ \text{kVA}$$

In three-phase circuits:

$$Kilovolt\text{-}amperes = \frac{volts \times amperes \times 1.73}{1000}$$

**Example:**

A three-phase generator, 480 volts, supplies a current of 40 amperes. Find the kVA rating.

**Solution:**

$$\text{kVA} = \frac{E \times I \times 1.73}{1000} = \frac{480 \times 40 \times 1.73}{1000} = 33.2\ \text{kVA}$$

## TRANSFORMER FORMULAS

A transformer is a device that transfers electrical energy from one circuit to another while changing the voltage. It consists of coils or windings around a magnetic core. The "input" side coil is called the "primary"; the "output" side coil is called the "secondary."

Figure 1.5

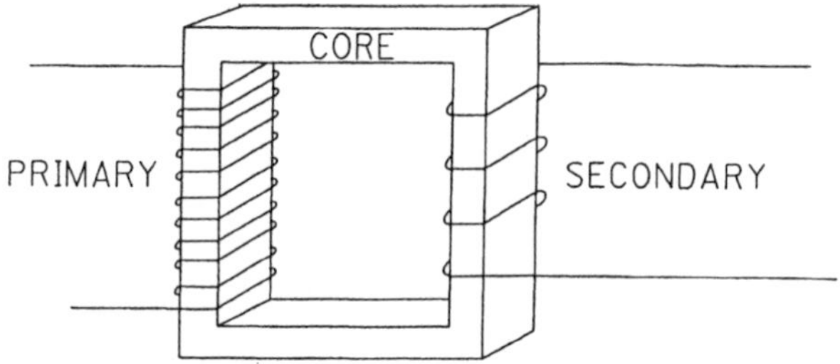


Unless employed by a power company, electricians will not be working on transformers. Even so, it is useful to understand the relationship of voltage input and voltage output. This relationship is stated in the "Transformer Rule" which is stated mathematically:

$$\frac{\textit{PRIMARY VOLTAGE}}{\textit{SECONDARY VOLTAGE}} = \frac{\textit{turns of primary}}{\textit{turns of secondary}}$$

Power output from a transformer is (ideally) the same as the power input. Since the transformer changes voltage, then it must also change current (power = volts x amperes). Amperage changes in proportion to the voltage but inversely. Mathematically, this is stated as:

$$\frac{primary\,voltage}{secondary\,amperage} = \frac{secondary\,voltage}{primary\,amperage}$$

Figure 1.6

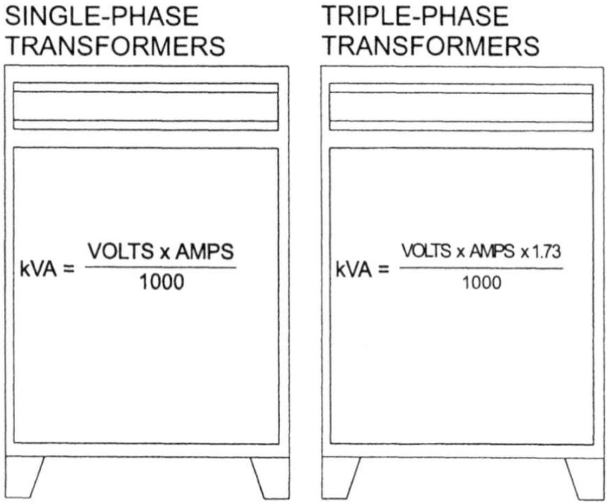


For all the formulas and examples that follow, the designations shall be as follows:

$E_P$ = primary voltage
$E_S$ = secondary voltage
$I_P$ = primary current
$I_S$ = secondary current
$t_P$ = number of turns in primary coil
$t_S$ = number of turns in secondary coil

<u>To find secondary current in a transformer</u>:

$$I_S = \frac{E_P \times I_P}{E_S} \qquad or \qquad I_S = \frac{I_S \times t_P}{t_S}$$

**Example 1:**

If a transformer has a primary current rating of 20 A at 460 V and boosts the secondary voltage to 4600 V, what current will the secondary provide?

**Solution 1:**

$$I_S = \frac{460 \times 20}{4600} = 2\,A$$

**Example 2:**

What current does the secondary of a transformer deliver if its primary draws 1 A, and has 500 turns in the primary coil and 100 turns in the secondary?

**Solution 2:**

$$I_S = \frac{1 \times 500}{100} = 5\,A$$

To find secondary voltage:

$$E_S = \frac{E_P \times I_P}{I_S} \text{ or } E_S = \frac{E_P \times t_S}{t_P}$$

**Example 1:**

A transfomer having a primary voltage of 440 volts draws 0.5 amperes at the primary winding and supplies 2 amperes to the secondary circuit. What is the secondary voltage?

**Solution 1:**

$$E_S = \frac{440 \times 0.5}{2} = 110\,V$$

**Example 2:**

What is the secondary voltage of a transformer with a primary voltage of 120 V, 400 turns in the primary coil, and 600 turns in the secondary coil?

**Solution 2:**

$$E_S = \frac{120 \times 600}{400} = 180\,V$$

## POWER FACTOR

If the average actual power is divided by the average volt-amperes, the fraction thus obtained (usually expressed as a decimal) is called the "power factor" of the circuit.
Stated another way, power factor is the ratio of the true power (in watts) to the apparent power (volt-amperes). This factor is never greater than 1.00.

Finding power factor in single-phase circuits:

$$Power\,factor = \frac{watts}{volt \times amperes}$$

**Example:**

What is the power factor of a device using 6000 watts operating on a 240 V AC circuit drawing 30 A?

**Solution:**

$$PF = \frac{W}{E \times I} = \frac{6000}{240 \times 30} = 0.833$$

Finding power factor for three-phase circuits:

$$Power\,factor = \frac{watts}{volts \times amperes \times 1.732}$$

**Example:**

What is the power factor of a device using 6 kilowatts operating on a 240 volt, three-phase circuit drawing 20 amperes?

**Solution:**

$$PF = \frac{Watts}{E \times I \times 1.732}$$

$$= \frac{6000}{240 \times 20 \times 1.732} = \frac{6000}{8313.6} = 0.722$$

## FINDING VOLTAGE DROP

Figure 1.7

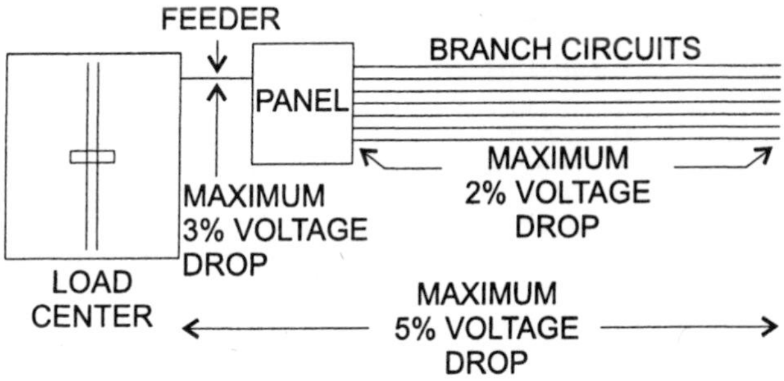

In single-phase circuits:

Voltage drop = 2 X resistivity of conductor X one-way length of circuit X current divided by the cross section of the conductor in circular mils.

$$Vd = \frac{2K \times L \times I}{D}$$

K, the resistivity of a conductor, is a constant. For copper conductors, K = 10.8; for aluminum, the constant is 17.

In three-phase circuits:

Voltage drop in a three-phase circuit is computed in the same way as in a single-phase circuit. However, the cross section of the conductor is multiplied by .866. The formula is as follows:

$$Vd = \frac{2K \times L \times I}{D \times 0.886}$$

## SIZING CONDUCTORS TO LIMIT VOLTAGE DROP

Voltage drop and proper wire sizing are key factors in planning circuits. Often, circuits with long conductors will have excessive voltage drop from improperly sized conductors which prevents the device from operating. Good design practice is to limit voltage drop to 3% but never more than 5%.

**For example:**

A remote stop-start station is located 150 ft. from the motor starter. The coil in the starter draws 7.2 amperes at 120 V AC. Size the conductor to limit voltage drop to the recommended 3%.

Using the voltage drop formula:

$$2K \times L \times I = CM$$

Where:

K = 10.7 constant for copper
L = Length of circuit (one way in feet)
I = Current drawn by device
Vd = Voltage drop
CM = Area of conductor in circular mils

Now, find the proper wire size to avoid excessive voltage drop by inserting the values into the formula:

$$\frac{214 \times 150 \times 7.2}{3.6} = 6420\,CM$$

To convert circular mils to American Wire Gauge sizes, refer to the following chart (See also Table 8 in the NEC):

| AWG | CIRCULAR MILS |
|---|---|
| 18 | 1620 |
| 16 | 2580 |
| 14 | 4110 |
| 12 | 6530 |
| 10 | 10380 |
| 8 | 16510 |
| 6 | 26240 |
| 4 | 41740 |

| AWG | CIRCULAR MILS |
|---|---|
| 3 | 52620 |
| 2 | 66360 |
| 1 | 83690 |
| 1/0 | 105600 |
| 2/0 | 133100 |
| 3/0 | 167800 |
| 4/0 | 211600 |

## INDUCTIVE REACTANCE (OHMS)

Inductive reactance is the counterelectromotive force that acts like a resistance that limits current in an AC circuit. Its symbol is $X_L$.

To find inductive reactance:

$$X_L = 2\pi fL$$

Where $f$ is the frequency of the AC circuit.

**Example:**

Determine the inductive reactance of a 4 henry coil in a 60 cycle AC circuit.

**Solution:**

$$X_L = 2 \times 3.14 \times 60 \times 4 = 1507 \, ohms$$

## CAPACITIVE REACTANCE

Capacitive reactance is the opposition to alternating current caused by capacitance.

To find capacitive reactance:

$$X_C = \frac{1}{2\pi fc}$$

**Example:**

What is the capacitive reactance of a 60 cycle circuit having a 6 microfarad capacitor?

**Solution:**

$$X_C = \frac{1}{2 \times 3.14 \times 60 \times 0.000006} = \frac{1}{0.0023} = 442.3 \, ohms$$

## CAPACITANCE

Capacitance, basically, is the ability of an electrical circuit, DC or AC, to store electric energy by means of an electrostatic field and, also, the ability to release this energy later. Devices that store the energy are called capacitors. This capacitance can be computed as follows:

<u>In a series circuit</u>:

Total capacitance is equal to the sum of the reciprocals of the capacitors in a series circuit.

$$\frac{1}{C_T} = \frac{1}{C_1} + \frac{1}{C_2} + \frac{1}{C_3} + etc.$$

### Example:

What would be the total capacitance of 4 capacitors in a series circuit, each having values of 16 microfarads?

### Solution:

$$\frac{1}{C_T} = \frac{1}{16} + \frac{1}{16} + \frac{1}{16} + \frac{1}{16} = \frac{4}{16} = 4 \text{ microfarads}$$

### Example (in parallel):

What is the total capacitance of 4 capacitors in parallel, each having values of 4 microfarads?

### Solution:

$$\begin{aligned} C_T &= C_1 + C_2 + C_3 + C_4 \\ &= 4 + 4 + 4 + 4 \\ &= 16 \text{ microfarads } (mf) \end{aligned}$$

## IMPEDANCE

Impedance (ohms) is the total opposition to the flow of current in a circuit. To find:

Impedance when current and voltage are known:

$$impedance\,(Z) = \frac{E\,(voltage)}{I\,(current)}$$

**Example:**

Find the impedance of an AC circuit with a current of 15 A and voltage of 150 V.

**Solution:**

$$Z = \frac{150}{15} = 10\,ohms$$

When resistance and reactance are known:

$$Z = \sqrt{R^2 + X^2}$$

**Example:**

Find the impedance of an AC circuit having a resistance of 7 ohms and a reactance of 8.

**Solution:**

$$Z = \sqrt{7^2 + 8^2} = \sqrt{49 + 64} = \sqrt{113} = 10.63\,ohms$$

When resistance and inductive and capacitive reactances are known:

$$Z = \sqrt{R^2 + (X_L - X_C)^2}$$

**Example:**

What is the impedance of a circuit having 12 ohms of resistance, 9 ohms inductive reactance, and 6 ohms capacitive reactance?

**Solution:**

$$Z = \sqrt{10^2 + (9-6)^2} = \sqrt{100+9} = \sqrt{109} = 10.44\ ohms$$

Figure 1.8

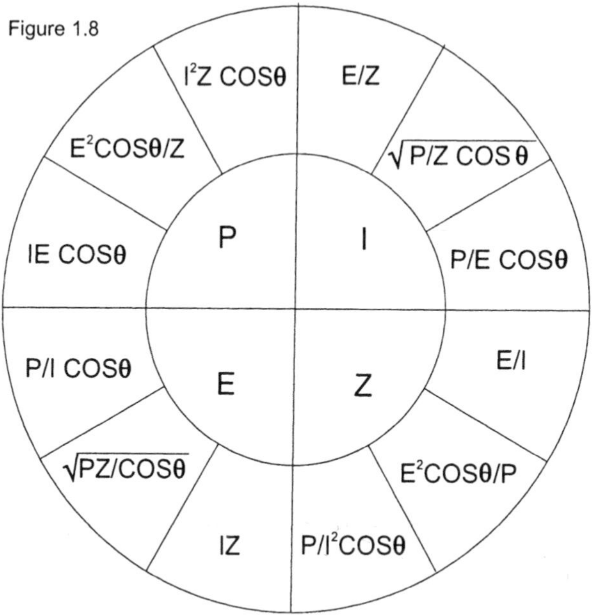


**This "wheel" includes the Ohm's Law formulas for alternating current formulas.**

| Summary of Formulas for Calculating Amperes, Horsepower, Kilowatts, and kVA in AC Current | | |
|---|---|---|
| TO FIND | ALTERNATING CURRENT | |
| | SINGLE PHASE | TWO-PHASE, FOUR- WIRE |
| AMPERES (HP KNOWN) | $\frac{HP \times 746}{E \times \%EFF \times PF}$ | $\frac{HP \times 746}{E \times \%EFF \times PF \times 2}$ |
| AMPERES (kW KNOWN) | $\frac{kW \times 1000}{E \times PF}$ | $\frac{kW \times 1000}{E \times PF \times 2}$ |
| AMPERES (kVA KNOWN) | $\frac{kVA \times 1000}{E}$ | $\frac{kVA \times 1000}{E \times 2}$ |
| KILOWATTS | $\frac{E \times I \times PF}{1000}$ | $\frac{E \times I \times PF \times 2}{1000}$ |
| KILOVOLT-AMPERES | $\frac{E \times I}{1000}$ | $\frac{E \times I \times 2}{1000}$ |
| HORSE-POWER | $\frac{E \times I \times \%EFF \times PF}{746}$ | $\frac{E \times I \times \%EFF \times PF \times 2}{746}$ |
| $PERCENT\ EFFICIENCY = \%EFF = \frac{OUTPUT}{INPUT}$ | | |
| $POWER\ FACTOR = PF = \frac{POWER\ USED\ (WATTS)}{APPARENT\ POWER} = \frac{kW}{kVA}$ | | |
| NOTE: | SINGLE-PHASE FORMULAS DO NOT USE ( 2 OR 1.73 ) | |
| | TWO-PHASE, FOUR-WIRE FORMULAS DO NOT USE ( 1.73 ) | |

# UNIT 2

# SERVICE ENTRANCE, FEEDERS, AND RELATED DEVICES

## INSTALLING SERVICE

The service entrance involves all the wires, devices, and fittings used between the power company's step-down transformer and the building it serves. All electrical power supplied to the building first passes through this service. The wires may be run overhead or underground.

Several important considerations in locating and running the service entrance are:

1. Whether in cable or conduit, wires should take as straight a path as possible and be kept as short as possible to avoid voltage drop

2. Wires should enter the building as close to the service panel as possible

3. A service disconnect should be at or near the service panel.

4. Service panels should be located for easy access near the major electrical appliances or electrical equipment of the building.

5. All service equipment should be protected from physical damage and exposure to moisture, dust, or other types of material that could affect operation.

6. Never place service equipment in bathrooms, closets, storerooms, or damp basements.

7. The meter enclosure (socket) should be mounted about 5 1/2 ft. above grade. Always check with the local utility.

Figure 2.1a

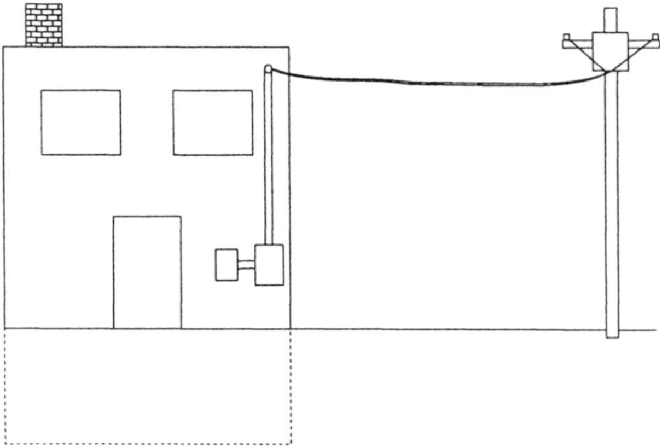

**Diagram of overhead service drop.**

Figure 2.1b

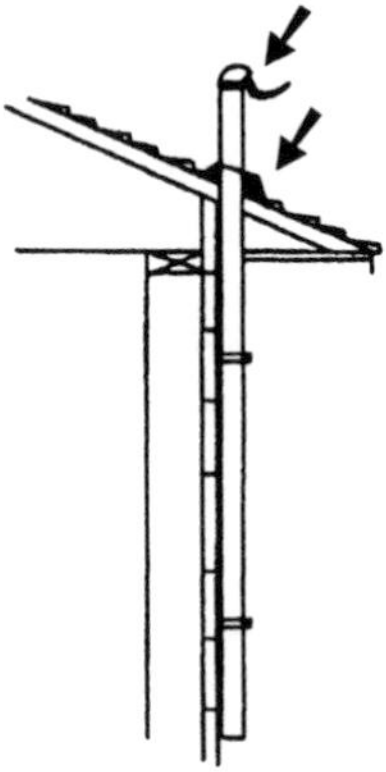

**Typical installation of a service riser when attached to a residence. (Meter and disconnect are not pictured.) Note rain boot and drip loop.**

For most applications, there shall be only one service for any building or other structure unless permitted in 230.2 (A)-(D).

Service drop cables containing or supporting several conductors shall be attached to buildings or to other structures by fittings identified for use with service conductors. Open conductors shall be attached to fittings identified for use with service conductors or to noncombustible, nonabsorbent insulators securely attached to the building or other structure.

Individual conductors shall be covered with an extruded thermoplastic or thermosetting insulation that will protect against detrimental current leakage, atmospheric or other conditions notwithstanding. Exception: the grounded conductor of a multiconductor cable shall be permitted to be bare.

Figure 2.2

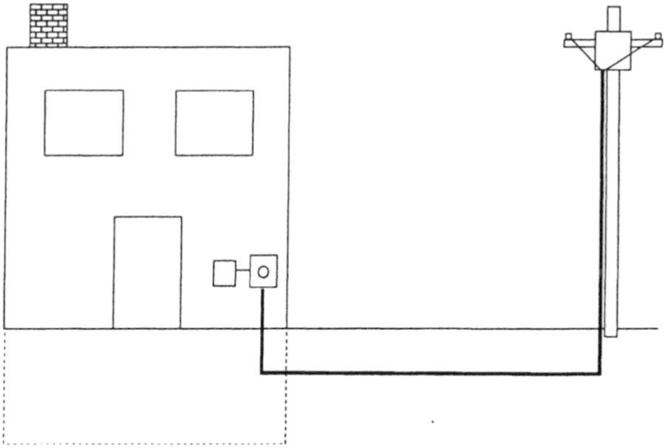

**Diagram of a service lateral.**

Conductors used in an underground service lateral shall withstand exposure to all conditions of exposure without detrimental current leakage. Such conductors shall also be insulated for the applied voltage. Exceptions: a grounded conductor may be uninsulated in the following situations:

1. Bare copper enclosed in a raceway.

2. Bare copper for direct burial where copper is considered to be suitable for the soil conditions.

3. Bare copper for direct burial with no regard to soil conditions where it is part of a cable assembly identified for underground use.

4. Aluminum or copper-clad aluminum with no individual insulation or covering where it is part of a cable assembly identified for underground use in a raceway or for direct burial.

5. Minimum size: not smaller than No. 8 copper or No. 6 aluminum or copper-clad aluminum.

6. Protection against damage: Refer to Table 300.5, NEC, or the chart, "Minimal Depths in Inches for Buried Conductors—0 to 600 V, Nominal."

Care should be taken in backfilling trenches containing underground cable service. Fill materials containing large rock, paving materials, cinders, any large or sharply angular substance, or corrosive materials shall NOT be used in places where these materials may cause damage to raceways, cables or other substructures or prevent proper compacting of the fill, or cause corrosion of raceways, cables, or other substructures.

Ferrous metal raceways, cable trays, cablebus, cable covering (armor), auxiliary gutters, boxes, cable sheathing, cabinets, metal elbows, couplings, fittings, supports, and support hardware shall be suitably protected against corrosion - inside and outside - (except threads at joints) by a coat of listed corrosion-resistant material. When enamel is used on such parts, they shall NOT be used out-of-doors or in wet locations such as laundries, dairies, or canneries. Boxes and cabinets that have an approved system of organic coatings and are marked "raintight," "rainproof," or "outdoor type" shall be permitted out-of-doors.

Table 2.1

| Minimal Depths in Inches for Buried Conductors: 0--600 V, Nominal | | | |
|---|---|---|---|
| (Includes conductors in cable, conduit or any type raceway) | | | |
| | Wiring Method or Circuit* | | |
| **Location** | **1** | **2** | **3** |
| All except those specified below | 24 | 6 | 18 |
| In a trench beneath 2 in. of concrete or its equivalent | 18 | 6 | 12 |
| Under a building | If in a raceway 0 | 0 | 0 |
| Beneath min. 4 in. of concrete ext. slab; no traffic; slab extending no less than 6 in. beyond underground installation | 18 | 4 | 4 |
| Under thoroughfares or parking lots | 24 | 24 | 24 |
| Driveways and parking areas of one- & two-family dwellings | 18 | 18 | 18 |
| In or under airport runways, including abutting or no trespassing areas | 18 | 18 | 18 |

* Key to numbered columns:
1 = Direct burial, cables or conductors.
2 = Protected by rigid or intermediate metal conduit.
3 = Nonmetallic raceways marked for direct burial w/o concrete encasement or other approved raceway.
NOTE: Shallower depths permitted where cables & conductors rise for terminations or splices or where access is required.
NOTE: Where a wiring method listed in cols. 1-3 is used in cols. 4 & 5, shallower burial depth is permitted.
NOTE: Where solid rock prevents such burial, all wiring must be encased in metal or nonmetallic raceways permitted for direct burial. Raceways shall be covered with a minimal 2 in. of concrete down to bedrock.
NOTE: Cover is defined as the distance in inches between the upper edge of the buried conductor, cable, conduit, or other raceway and the surface of finished grade, concrete, or similar cover.
Please see NEC Table 300.5 for additional information.

Table 2.1 (cont.)

| Minimal Depths in Inches for Buried Conductors: 0-600 V, Nominal | | |
|---|---|---|
| (Includes conductors in cable, conduit or any type raceway) | | |
| | Wiring Method or Circuit* | |
| Location | 4 | 5 |
| All except those specified below | 12 | 6 |
| In a trench beneath 2 in. of concrete or its equivalent | 6 | 6 |
| Under a building | If in a raceway 0 | If in a raceway 0 |
| Beneath min. 4 in. of concrete ext. slab; no traffic; slab extending no less than 6 in. beyond underground installation | 6 -- Direct Burial<br>4 -- Raceway | 6 -- Direct Burial<br>4 -- Raceway |
| Under thoroughfares or parking lots | 24 | 24 |
| Driveways and parking areas of one- & two-family dwellings | 12 | 18 |
| In or under airport runways, including abutting or no trespassing areas | 18 | 18 |

* Key to numbered columns:
4 = Residential branch circuits rated at 120 V with GFCI protection & max. overcurrent protection of 20 A.
5 = Circuits controlling irrigation & landscaping lights of no more than 30 V & installed with Type UF cable or encased in other identified cable or raceway.
NOTE: Shallower depths permitted where cables & conductors rise for terminations or splices or where access is required.
NOTE: Where a wiring method listed in cols. 1-3 is used in cols. 4 & 5, shallower burial depth is permitted.
NOTE: Where solid rock prevents such burial, all wiring must be encased in metal or nonmetallic raceways permitted for direct burial. Raceways shall be covered with a minimal 2 in. of concrete down to bedrock.
NOTE: Cover is defined as the distance in inches between the upper edge of the buried conductor, cable, conduit, or other raceway and the surface of finished grade, concrete, or similar cover.
Please see NEC Table 300.5 for additional information.

Figure 2.3

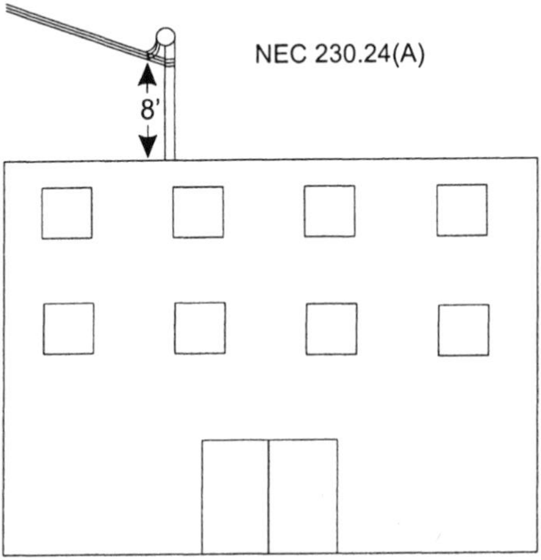


**Service drop clearances at all points on a flat roof shall be not less than 8 ft.**

Figure 2.4

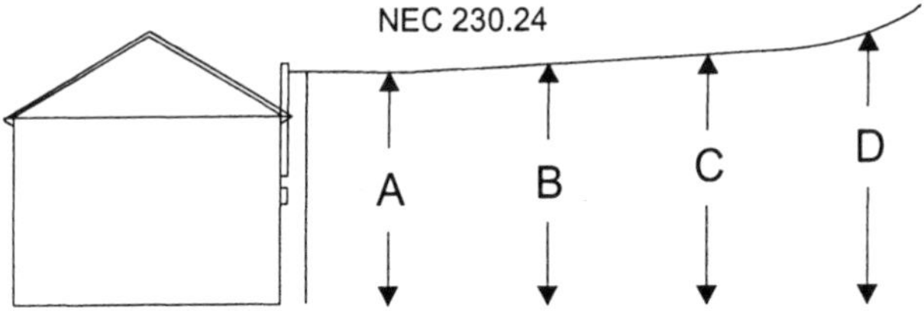


**For service drops, the Electrical Code specifies the following vertical clearances (next page):**

A. Ten ft. over steps, porches, decks, and sidewalks with service not more than 150 V to ground.
B. Twelve ft. over property driveways having no truck traffic and with service not over 300 V to ground.
C. Fifteen ft. over property driveways having no truck traffic and when service is rated over 300 V to ground.
D. Eighteen ft. over public streets, roads, alleys, and parking lots and other locations where there is truck traffic.

**Rules for overhead service masts.** (Figures 2.5a-2.5d)

Figure 2.5a

**Drip loops, designed to keep water from entering the weatherhead, are optional when the service drop is located below the weatherhead.**

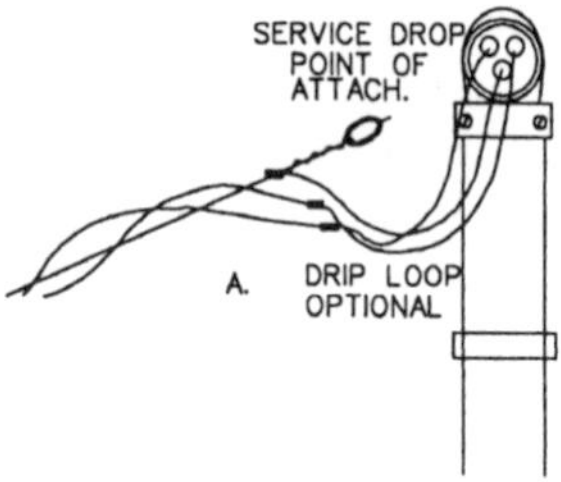


Figure 2.5b

**However, a drip loop is necessary wherever the point of attachment of the service drop is above the weatherhead. In no case may the point of attachment be more than 2 ft. above the weatherhead. Service conductors should extend outward through the head about 3 ft. to form a suitable drip loop.**

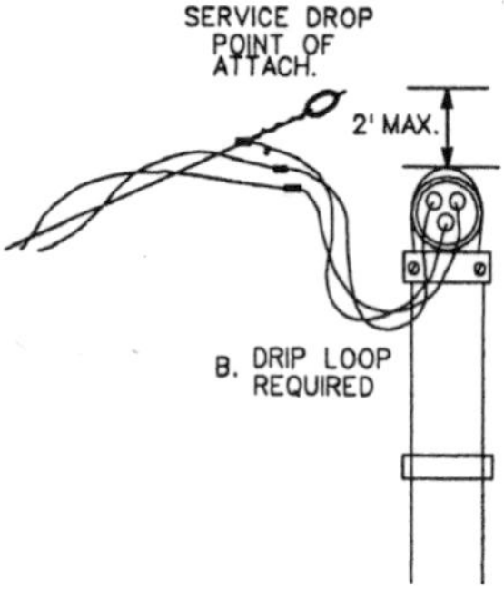

Figure 2.5c

**Service drop attachment must be below the service head in a gooseneck arrangement.**

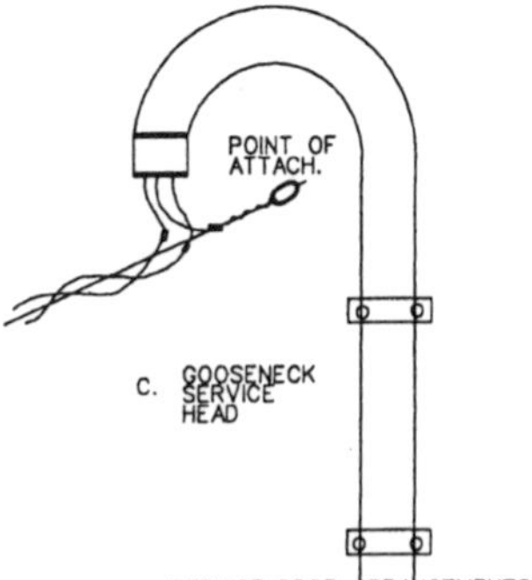


Figure 2.5d

**Support strapping must be installed on cable entrances following these recommended intervals.**

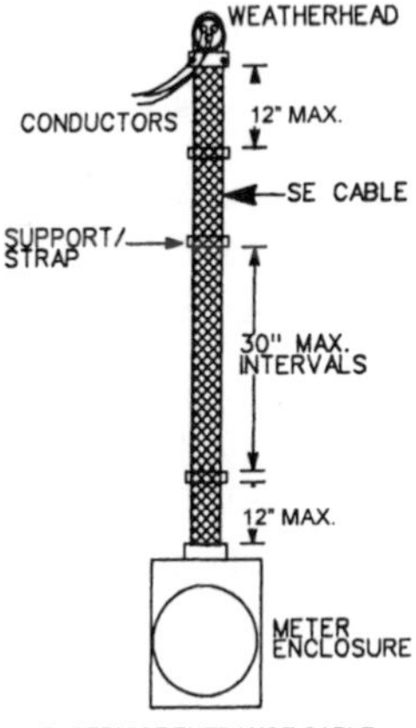


D. SERVICE ENTRANCE CABLE SUPPORT REQUIREMENTS

Figure 2.6

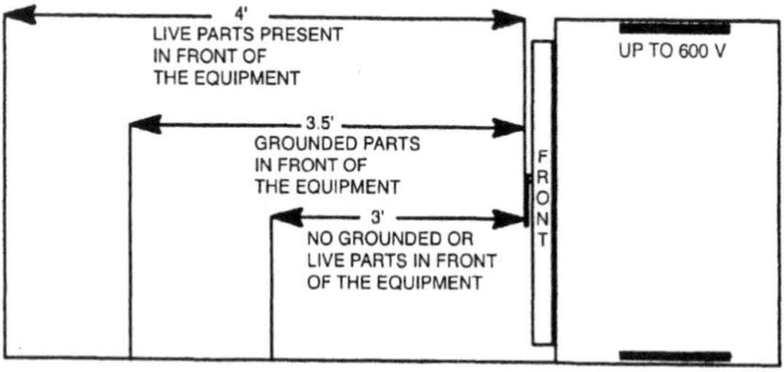


**Required working clearance distance in front of electrical equipment. See Conditions, NEC Table 110.26 (A)(1)**

The National Electrical Code 110.26 states that "sufficient access and working space shall be provided and maintained" around all electrical equipment for safe, easy operation and maintenance. Distances are to be measured from the live parts, if they are exposed, or from the enclosure front or opening, if equipment is enclosed. Refer also to NEC: 110.31, 110.32, 110.33, and 110.34.

## SIZING SERVICE ENTRANCE CONDUCTORS

Ampacity: The current-carrying ability of conductors shall not be less than required to supply loads as computed in the National Electrical Code, 220:

Part III - Feeders and Services.

Part IV - Optional Calculations for Computing Feeders & Service Loads.

Part V - Method for Computing Farm Loads: For specified circuits, ampacity of feeder conductors shall not be less than 30 A where the load consists of any of the following:

1. Two or more two-wire branch circuits supplied by a two-wire feeder.

2. More than two two-wire branch circuits supplied by a three-wire feeder.

3. Two or more three-wire branch circuits supplied by a three-wire feeder.

4. Two or more four-wire branch circuits supplied by a three-phase four-wire feeder - The feeder conductor ampacity shall not be less than that of the service entrance conductors where the feeder conductors carry the load supplied by service entrance conductors with an ampacity of 55 A or less. (See NEC, Article 215).

Table 2.2

| Steps for Sizing Service Entrance Conductors | |
|---|---|
| **Procedure** | **NEC Code Reference** |
| 1. Determine lighting load. | Table 220.12 |
| 2. Find small appliance load for kitchen and laundry. | 220.14 |
| 3. Add totals from Steps 1 and 2. Add in Code demand factors. | Table 220.42<br>Table 310.16 |
| 4. Compute and add in loads for special and fixed appliances: a) Heating @ 100% demand. b) Dryer @ 100% demand. c) Range (Code allows demand factor based on size & number). Add demand factor to total. | 220.51<br>220.54<br>Table 220.55 |
| 5. Add up all demands loads from the previous steps. | |
| 6. Divide the demand load total by voltage supplied to get needed carrying capacity from hot legs of the service entrance. Check NEC for sizes needed | Table 310.16 |
| See NEC Annex D for examples. | |

Table 2.2 (cont.)

| To calculate size of service neutral feeder: | |
|---|---|
| **Procedure** | **NEC Code Reference** |
| 1. Add demand loads of all 120 V items: a) General lighting, kitchen, laundry. b) Dishwasher @ 100% load. c) Clothes washer @ 75% load. | Table 220.42<br>220.12<br>220.52 & 220.53<br>220.61 |
| 2. List devices using 240 V, along with their demand loads. a) Range @ 70% load. b) Electric dryer @ 70% load. | Table 220.54<br>Table 220.55<br>220.61 |
| 3. Add results of Steps 1 & 2; divide by 240 V to find load the neutral must carry. Check NEC table for conductor size required. | Table 310.16 |
| 4. Compute and add in loads for special and fixed appliances: a) Heating @ 100% demand. b) Dryer @ 100% demand. c) Range (Code allows demand factor based on size & number) Add demand factor to total. | 220.51<br>220.54<br>Table 220.55 |

## UNDERSTANDING GROUNDING ELECTRODE REQUIREMENTS OF ELECTRICAL POWER SYSTEMS

A vital part of every electrical power distribution system is proper grounding. There are two distinct types:

1. EQUIPMENT GROUND. This is simply the connection to the system of all metal electrical equipment. It involves the third bare or green conductor in the circuit. All electrical equipment, conductor enclosures, as well as any nonelectrical equipment associated with the system, are bonded together. This assures an uninterrupted electrically conductive path to ground from any part of the system. The result is that all exposed conducting surfaces are at the same electrical potential. Thus, someone touching any two metal surfaces or any grounded metal surface, while in contact with the earth, would not experience a shock (difference in potential).

2. A GROUNDED CONDUCTOR. In this type of grounding, one of the circuit conductors, sometimes called the neutral, is grounded to keep the system at a fixed voltage in relation to ground. The grounding should be done at the source. The diagram following shows this grounding both at the transformer and at the service disribution panel in the building. Should a downstream equipment ground be connected to the grounded conductor it would make current-carrying conductors out of circuit components never intended for that purpose. A grounded conductor is normally insulated because it usually runs at some level of voltage above ground.

In a grounded system, a common grounding electrode conductor connects both the equipment ground and the grounded conductor to the grounding electrode. The equipment grounding conductor must be connected to the grounded conductor inside the service entrance equipment. In this way, the equipment grounding conductor has the lowest practical impedance. The common grounding electrode conductor is then connected to the grounded conductor at any convenient point on the supply side of the service disconnecting means. (See 250.24 and 250.26 of the National Electrical Code and to the following diagram.)

Figure 2.7

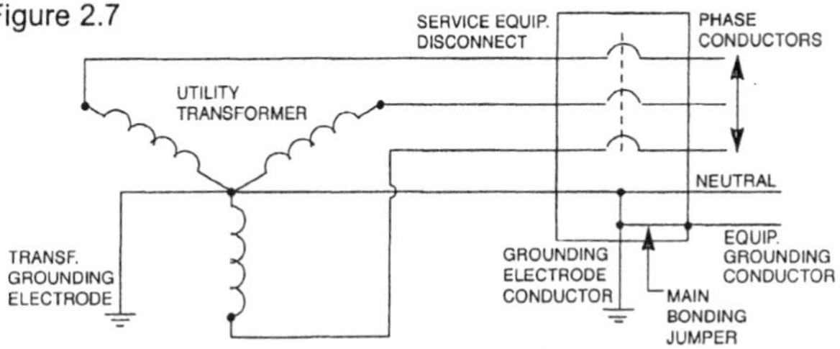


**System grounding of grounded neutral conductor is done at transformer and service panel. Note main bonding jumper bonds grounded neutral conductor to box, grounding electrode conductor, and equipment grounding conductor. The electrode is usually a copper rod driven to a depth of 8 feet. However, metal water pipes are a suitable electrode.**

Table 2.3

| GROUND ELECTRODE CONDUCTOR SIZE REQUIREMENTS OF AC SERVICES | | | |
|---|---|---|---|
| Sizes of Largest Service Entrance Conductor | | Size of Grounding Electrode Conductor | |
| Copper (Cu) | Aluminum (Al) or Cu Clad | Cu | Al or Cu Clad Al |
| *AWG or kcmil* | *AWG or kcmil* | *AWG* | *AWG or kcmil* |
| 2 or smaller | 1/0 or smaller | 8 | 6 |
| 1 or 1/0 | 2/0 or 3/0 | 6 | 4 |
| 2/0 or 3/0 | 4/0 or 250 kcmil | 4 | 2 |
| Over 3/0 thru 350 kcmil | Over 250 kcmil thru 500 kcmil | 2 | 1/0 |
| Over 350 kcmil thru 600 kcmil | Over 500 kcmil thru 900 kcmil | 1/0 | 3/0 |
| Over 6000 kcmil thru 1100 kcmil | Over 900 kcmil thru 1750 kcmil | 2/0 | 4/0 |
| Over 1100 kcmil | Over 1750 kcmil | 3/0 | 250 kcmil |
| See NEC Table 250.66 | | | |

When multiple sets of service-entrance conductors are used (allowed in NEC, 230.40, Exceptions), then the equivalent size of the service-entrance conductor shall be determined by the largest sum of the areas of the corresponding conductors of each set.

When there are no service-entrance conductors, the size of the grounding electrode conductor shall be determined by the equivalent size of the largest service-entrance conductor required for the load to be served.

Table 2.4

| Ampacities and Sizes of RHH, RHW, RHW-2, THHN, THHW, THW, THW-2, THWN, THWN-2, XHHW, XHHW-2, SE, USE, & USE-2 Conductors* | | |
|---|---|---|
| Copper AWG/kcmil | Aluminum and Copper-clad Aluminum (AWG/kcmil) | Service or Feeder Rating in Amperes |
| 4 | 2 | 100 |
| 3 | 1 | 110 |
| 2 | 1/0 | 125 |
| 1 | 2/0 | 150 |
| 1/0 | 3/0 | 175 |
| 2/0 | 4/0 | 200 |
| 3/0 | 250 kcmil | 225 |
| 4/0 | 300 kcmil | 250 |
| 250 kcmil | 350 kcmil | 300 |
| 350 kcmil | 500 kcmil | 350 |
| 400 kcmil | 600 kcmil | 400 |

* Conductor requirements listed here are for dwelling units using 120/240 voltage. They shall be permitted as 3-wire, single-phase service entrance conductors, service lateral conductors, and feeder conductors that serve as main power feeds to a dwelling unit. For an explanation of type (by letter designation), go to Table 310-13. See Table & Section 310.15(B)(6)

Figure 2.8a

**A - 120/240 V or 120/208 V single-phase 3-wire**

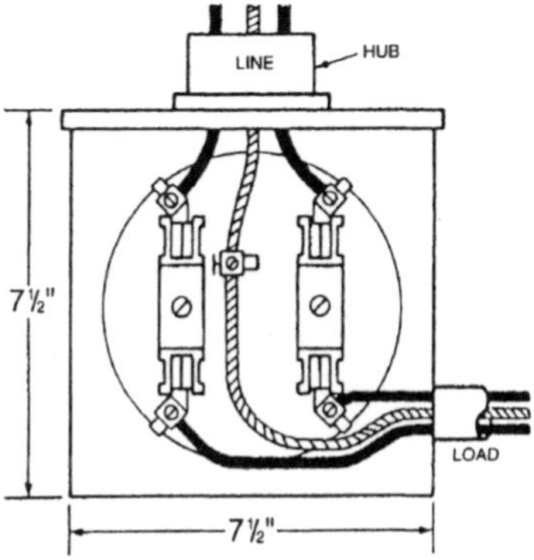

Figure 2.8b

**B - Typical wiring diagram for 100 to 150 A meter socket. Circuit is 120/240 V or 120/208 V single-phase 3-wire.**

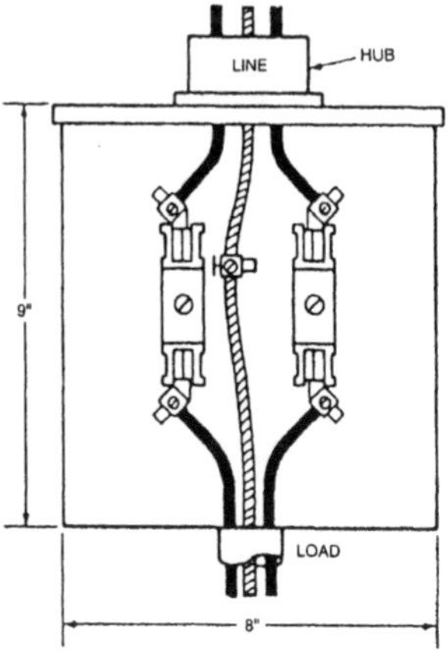


Figure 2.8c

**C - 120/240 V or 120/208 V single-phase, 3-wire.**

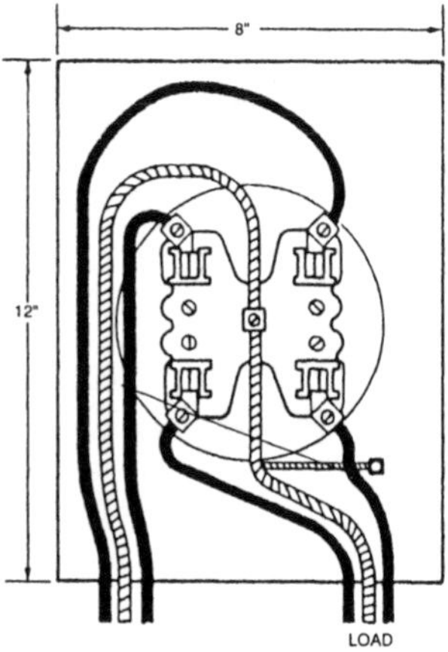


**The preceding diagrams are typical meter socket arrangements used for overhead and underground services.**

Figure 2.9

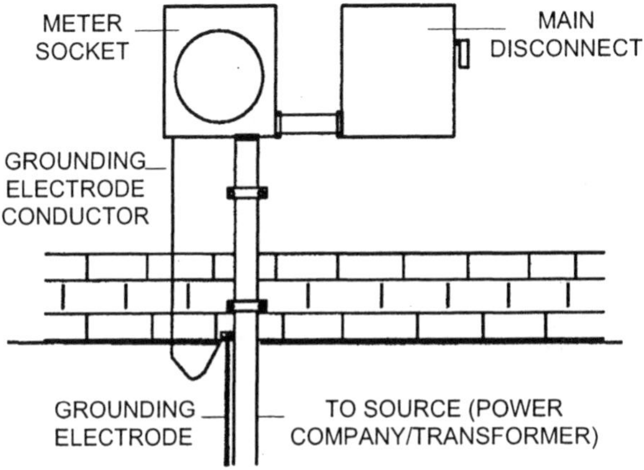


**Typical service lateral arrangement with exterior mounted main disconnect.**

## MULTIPLE OCCUPANCY SERVICE ENTRANCE

Multiple occupancy buildings (such as apartment buildings and commercial structures) may have any number of service-entrance conductor sets tapped from a single service drop or service lateral. The NEC permits up to six disconnects. The illustration following shows an arrangement with four disconnects. The taps are housed in the gutter.

Figure 2.10

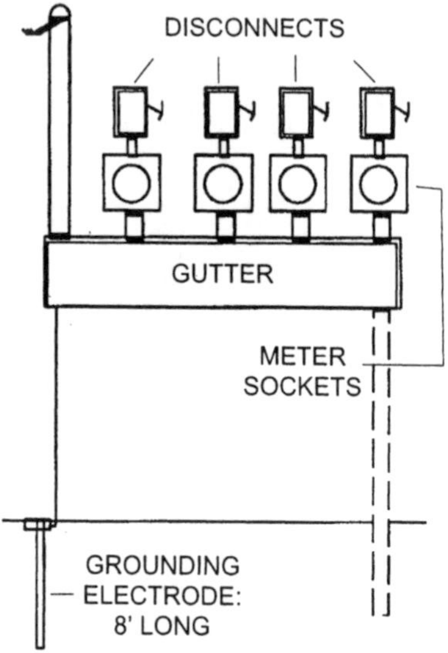


**Typical of an installation for a multiple occupancy building. Either a service drop or service lateral brings service to the gutter. Up to six disconnects are permitted.**

Figure 2.11

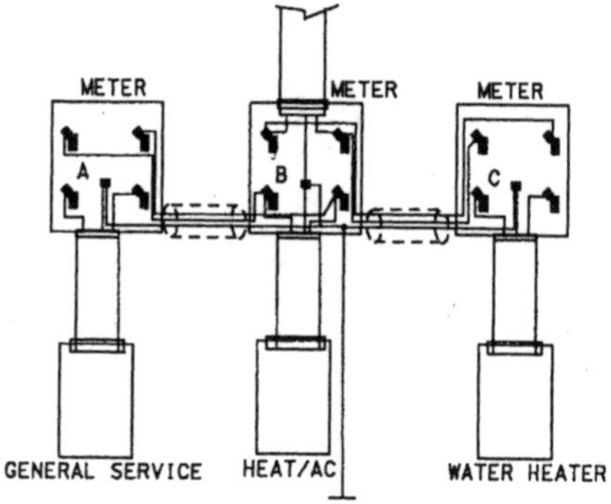


**This arrangement allows for separate metering of heating/AC and water heating.**

Figure 2.12

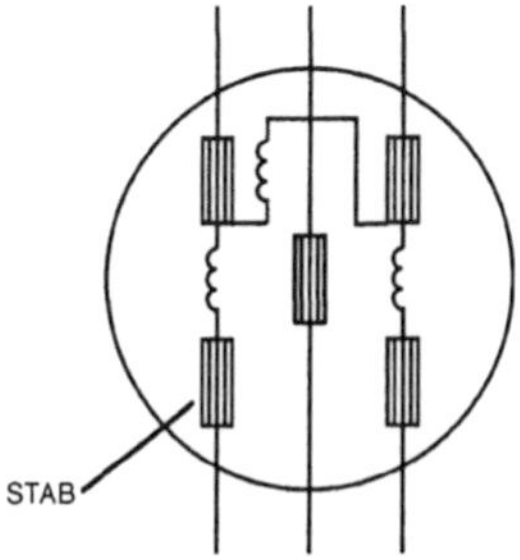


**Internal connections in a kilowatt-hour meter are slotted to receive stabs on the back of the meter. These stabs slide into the connections (lugs) shown above.**

## MULTIPHASE SYSTEMS

Commercial buildings, such as apartment buildings, hotels, motels, garages, factories and industrial parks, have heavier load requirements because of electrical motors and higher levels of lighting. In such cases, three-phase, four-wire systems have been developed.

Three-phase, four-wire systems, supply electric power in two service ratings: 120/208 V and 120/240 V, the last being the most commonly used. It provides both 120 volts for general lighting and receptacles, and 240 volts for heavy-duty appliances, welders, large electric motors and other electrical equipment.

The three illustrations following show two schematics of a delta wired three-phase, four-wire system. The third shows how the system is wired at the meter socket.

Figure 2.13

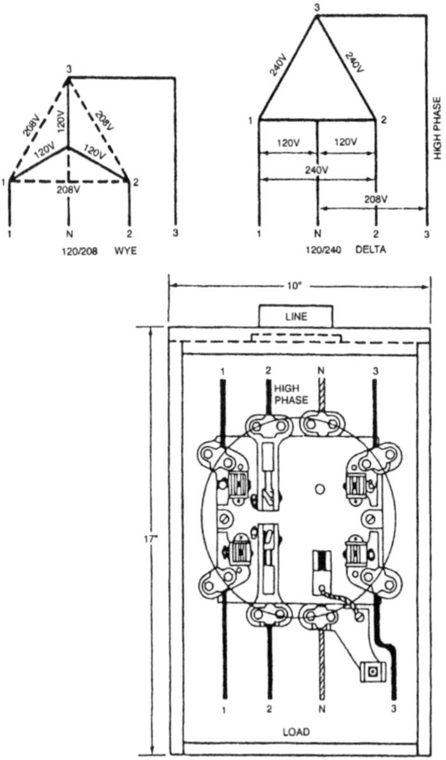


**Typical wiring arrangement for a delta and wye three-phase, 200 A service. The high phase is identified using a tag or color (usually orange) at each termination point. See NEC 110.15**

Figure 2.14

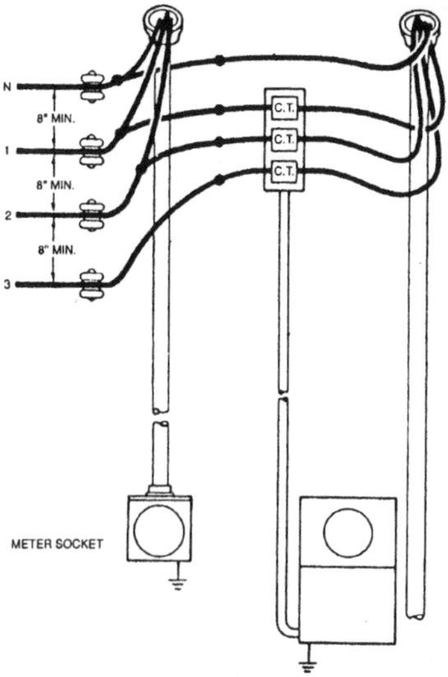


**Typical installation of a four-wire, three-phase, 120/240 volt or 120/208 volt system. One is self contained; the other (at right) is a current transformer meter installation.**

## WIRE-BENDING SPACE

NEC specifies (See Article 312) bending space of conductors in a cabinet or gutter. See illustration and table following:

Figure 2.15

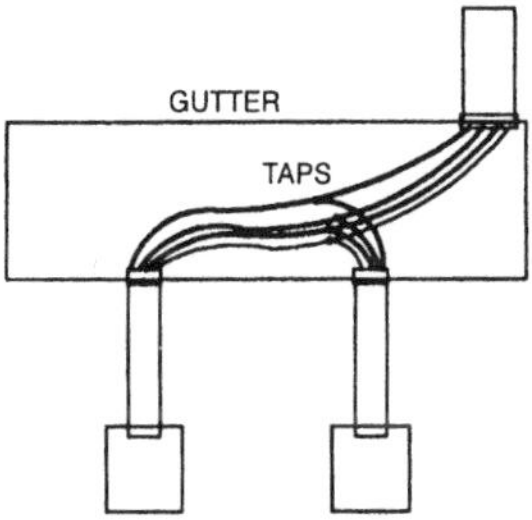

Table 2.5

| Minimum Wire-bending Space in Inches at Terminals and Minimum Width of Wiring Gutters | | | | | |
|---|---|---|---|---|---|
| **Size of Wire** | **Wires each Terminal** | | | | |
| (AWG or kcmil) | **1** | **2** | **3** | **4** | **5** |
| 14-10 AWG | Not Specified | - | - | - | - |
| 8-6 AWG | 1 1/2 | - | - | - | - |
| 4-3 AWG | 2 | - | - | - | - |
| 2 AWG | 2 1/2 | - | - | - | - |
| 1 AWG | 3 | - | - | - | - |
| 1/0-2/0 AWG | 3 1/2 | 5 | 7 | - | - |
| 3/0-4/0 AWG | 4 | 6 | 8 | - | - |
| 250 kcmil | 4 1/2 | 6 | 8 | 10 | - |
| 300-350 kcmil | 5 | 8 | 10 | 12 | - |
| 400-500 kcmil | 6 | 8 | 10 | 12 | 14 |
| 600-700 kcmil | 8 | 10 | 12 | 14 | 16 |
| 750-900 kcmil | 8 | 12 | 14 | 16 | 18 |
| 1000-1250 kcmil | 10 | - | - | - | - |
| 1500-2000 kcmil | 12 | - | - | - | - |
| Measure space in a straight line from end of lug or wire connector (in the direction that wire leaves the terminal) to wall, barrier, or obstruction. NOTE: To convert table to millimeters, multiply inches by 25.4. See NEC, Article 312. | | | | | |

Figure 2.16

**Code allows temporary service entrance installation during construction. Power outlets are often below or within the disconnect enclosure.**

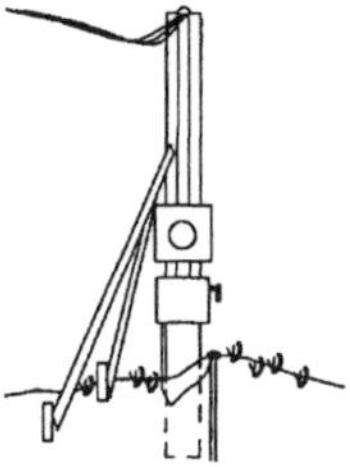

# UNIT 3

# BRANCH CIRCUITS

## BASIC CIRCUIT THEORY

Every circuit is defined and recognized by the path or paths of the current allowed to flow through it. Every circuit must have a switch for control of current, a fuse or circuit breaker to protect conductors and other devices from overcurrent damage and fire, and a "load" that allows the electricity to do work. This load may be a light, heater, or some other appliance, or a motor that drives equipment. See the circuit illustration below.

Figure 3.1

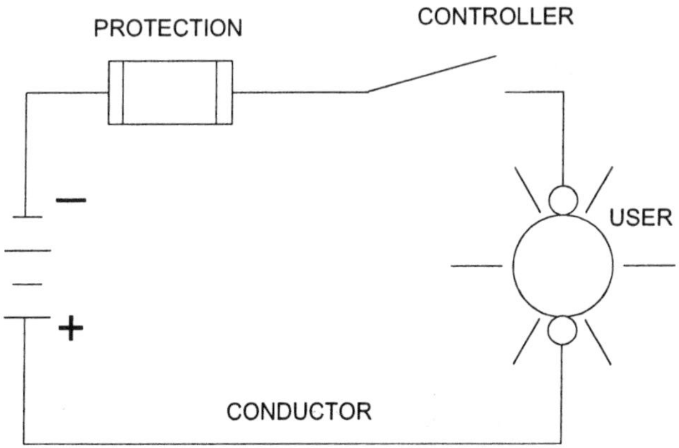


**The simplest form of an electrical circuit. The power source is a battery. The load is a light.**

## Series Circuits

Series circuits (see illustration) have but one path for current. They may have more than one energy-using device. It is a characteristic of all such circuits that when any part of the circuit interrupts current, the entire circuit is "dead" (open). Fuses and circuit breakers are always installed in series so that the circuit is protected. Another example of a series installation is a switch. It is always series-wired to control the circuit. (Not all circuits have switches. For example, receptacles, that must always be "hot" are not switched.)

Figure 3.2

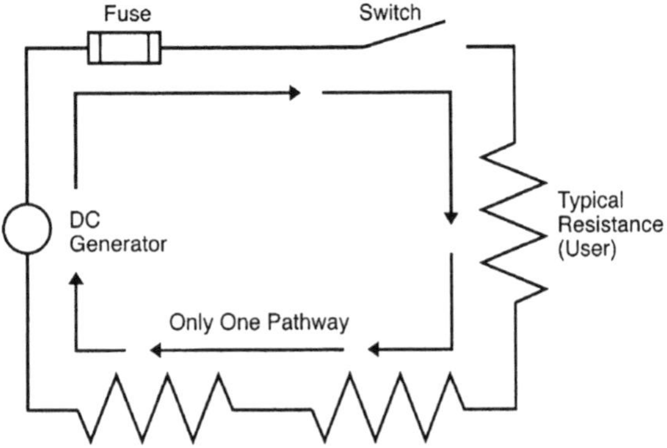


**Series circuits are "dead" (not operating) if any device in it is not conducting current.**

## Parallel Circuits

In parallel circuits current is "split" between two or more branches. It is clear that the current in any branch is less than the current of the power source. Total current is the sum of all the branch currents. Voltage across each branch load is the same as the total voltage from source.

Figure 3.3

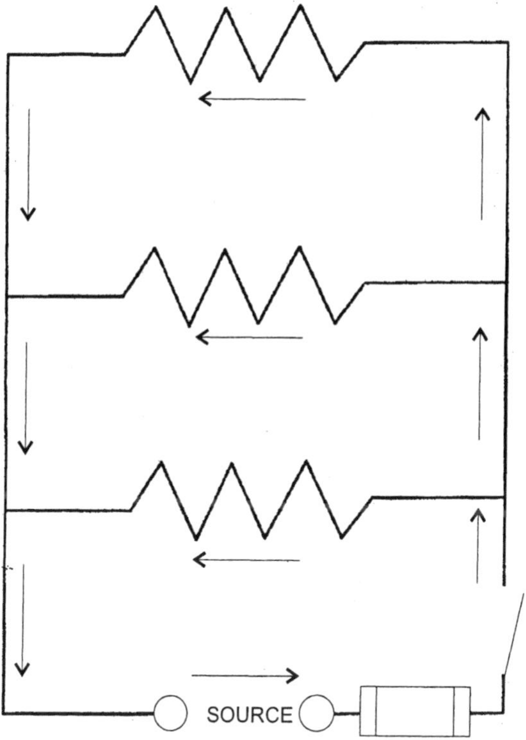


**In schematics, circuit devices are often represented by symbols. In this schematic drawing, the loads in the parallel paths are represented by the symbol for a resistance.**

## Series-Parallel (Compound) Circuits

Also called combination circuits, these circuits have some parts in series and some parts in parallel. Circuitry in buildings is of this type. In fact, most practical circuits are of this type.

To find the voltages, currents, and resistances in a series-parallel circuit, simply follow the rules of series circuits for the series part and parallel-circuit rules for the parallel part. For example: in a series circuit, the current is the same at all points. In a parallel circuit, there are one or more points where the current divides and flows in different paths or branches.

To determine if a circuit is series, parallel, or series-parallel, it is easiest to start at one terminal of the power source and trace the path of current through the circuit back to source at the other terminal. It is helpful, when trying to determine values (current, voltage and resistance) to redraw the circuit as shown in illustrations 3.4a and 3.4b below and on the following page. Then branches and loads are more easily recognized.

Figure 3.4a

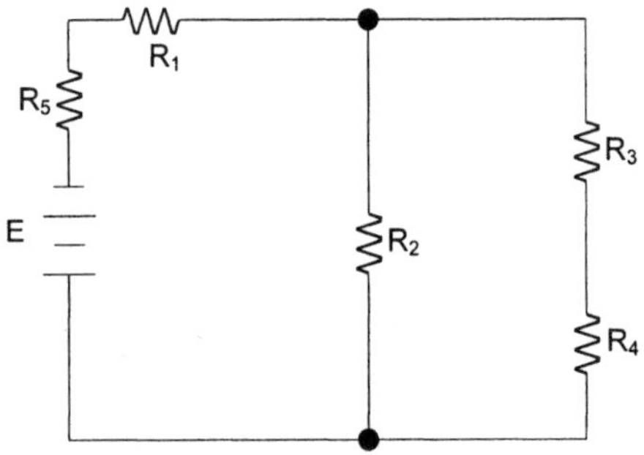

Figure. 3.4b

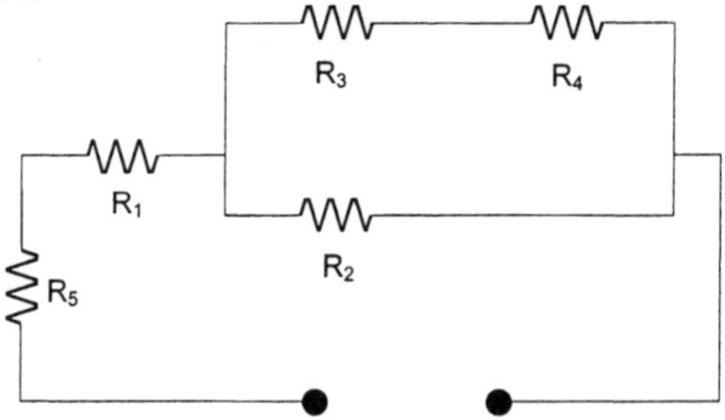


**Series-parallel circuits, also called "compound circuits," can be drawn like "B" (above) to make determining values easier.**

## SYSTEM GROUNDING AND BONDING REQUIREMENTS

System grounding is intentionally connecting one conductor of an electrical system to the earth. It is required of all building wiring systems. Usually the ground connection is made to the grounded neutral conductor if there is one. In a three-wire, three-phase delta system or a two-wire, single-phase 120 volt system it can be a line conductor.

Bonding ensures a continuous metallic, current-carrying path throughout a circuit's grounding system. Article 250 NEC requires bonding at or across:

1. All conduit connections in electrical service equipment.
2. Parts of an electrical system where a nonconducting material is used that would break continuity. Such bonding is accomplished by connecting a jumper around such nonconducting material.
3. All service equipment enclosures, such as boxes, whether outside or in a building.

Following schematics illustrate various aspects of grounding requirements.

Figure 3.5

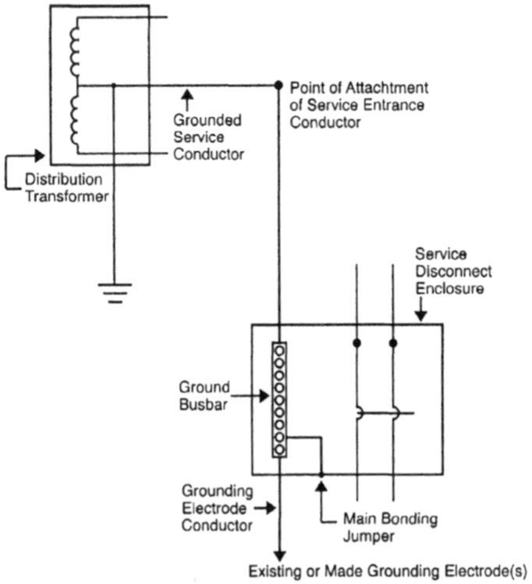


**Schematic of proper grounding of the neutral conductor from a transformer pole or pad to a service disconnect enclosure or service entrance panel.**

Figure 3.6

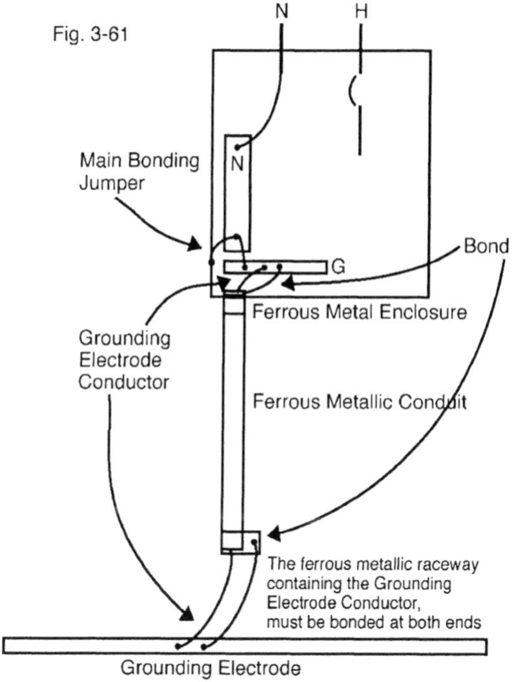


**Proper method of bonding a ferrous metallic raceway.**

Figure 3.7

Fig. 3.71

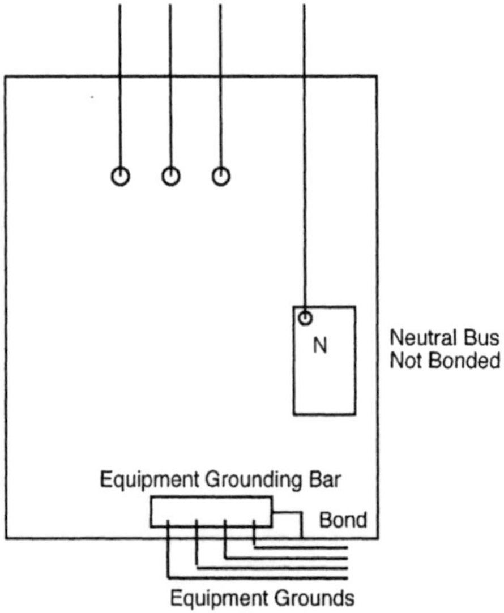


**IMPORTANT:**

1. The grounded conductor (neutral) may not be used as the equipment Ground in the load side of the service.
2. The Grounded Conductor shall be connected to the Grounding Electrode Conductor at the service and only at the service.

Figure 3.8

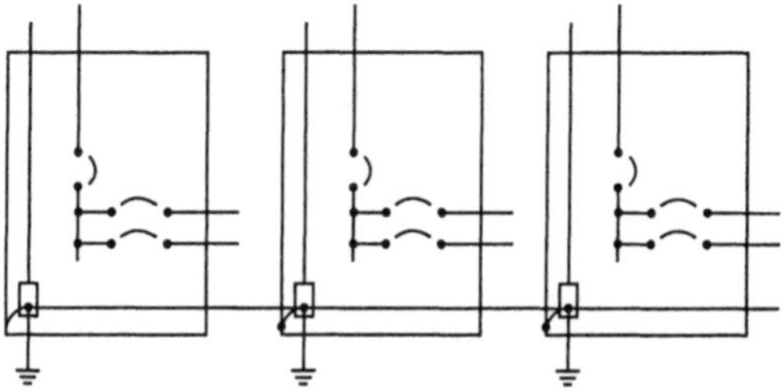

**Where a grounded service supplies more than one building, the grounding must be present at each building and be interconnected.**

Figure 3.9

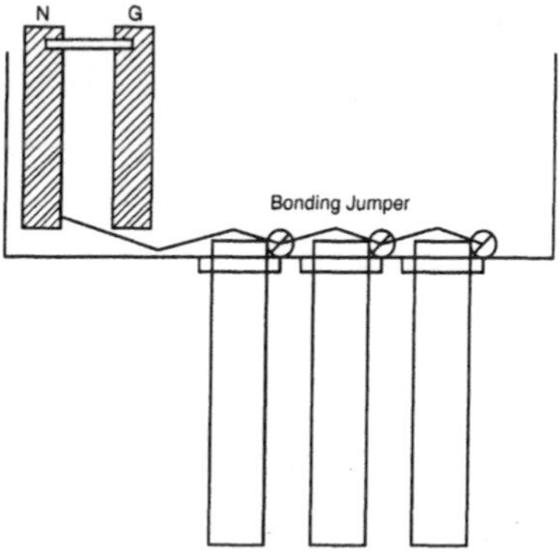


**Bonding jumper arrangement connects grounded conductor (neutral) to conduit of branch circuits coming out of a service panel.**

Figure 3.10

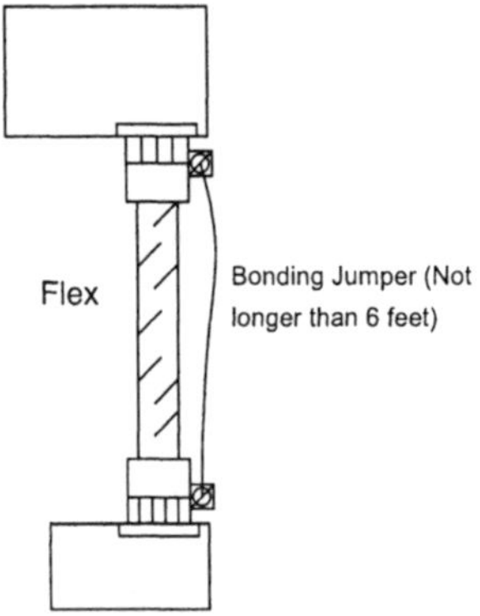


**This bonding jumper arrangement provides continuity between two metal boxes connected by nonmetallic conduit.**

Figure 3.11

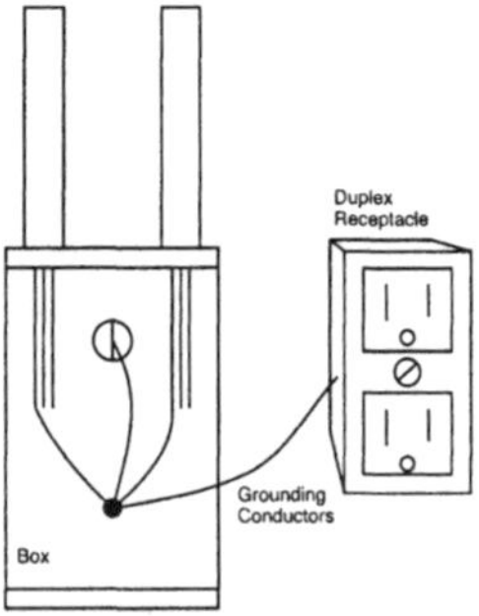


**Proper bonding of a receptacle with equipment ground. Green or bare grounding conductor bonds the receptacle and any electrical equipment to the metal box.**

## FINDING BRANCH LOADS FOR RESIDENTIAL ELECTRICAL WORK

The National Electrical Code specifies that the number of lighting circuits is to be determined as 3 volt-amperes for every square foot of occupied residential space. No specification is given for the number of receptacles. The Code says that the maximum load of any branch circuit is not to be greater than the circuit's rating. Rather than load a circuit to 100% capacity, a good rule to follow is:

$$\textbf{\textit{Circuit amperage}} \times \textbf{\textit{circuit voltage}} \times \mathbf{0.80}$$

For example, a 20 A branch circuit can carry a maximum load of:

$$20\ A \times 120\ V \times 0.80 = 1920\ VA$$

While this gives the total wattage for the circuit, there are variables to consider. Some outlets will carry heavy loads; others may only have a night light. As a general rule then, no outlet, whether single, duplex, or triple, should be rated at more than 180 W (VA). For example, suppose that a circuit will have two 150 W ceiling lights and four receptacles. What size branch circuit is required?

$$\begin{aligned} \textbf{\textit{2 ceiling fixtures @ 150 each...}}\ & \mathbf{300}\ \textbf{\textit{VA}} \\ \textbf{\textit{4 receptacles @ 180 each...}}\ & \underline{\mathbf{720}}\ \textbf{\textit{VA}} \\ & \mathbf{1020}\ \textbf{\textit{VA}} \end{aligned}$$

A 15 A circuit (15 X 120 X 0.80 = 1440) is indicated.

Figure 3.12

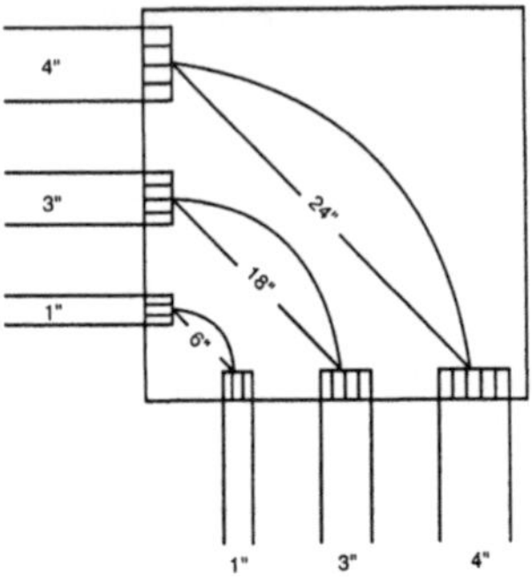


Correctly sizing angle pulls.

**Minimum distance required in an electrical box for making angle pulls is equal to six times the size of the conduit.**

Figure 3.13

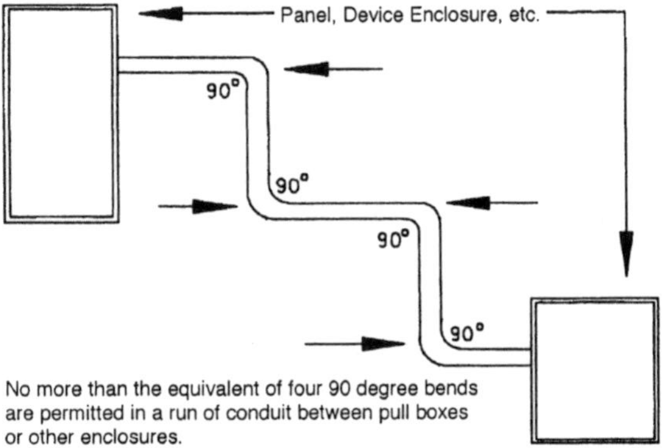


No more than the equivalent of four 90 degree bends are permitted in a run of conduit between pull boxes or other enclosures.

**The Code restricts number of 90-degree bends in conduit runs.**

Figure 3.14

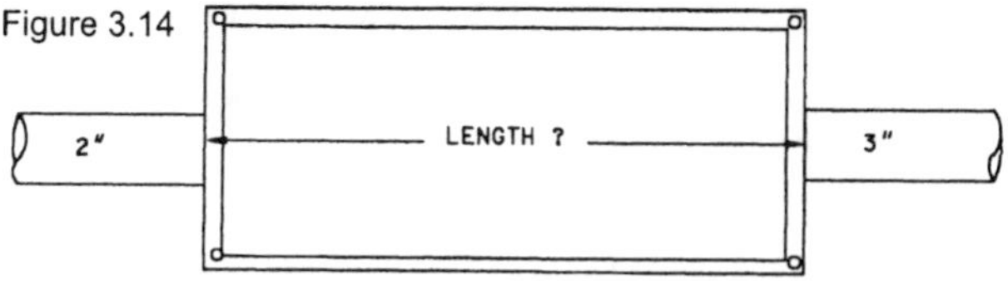


Pull box sizing: for straight pulls, the length of the box must not be less than 8x the trade diameter of the largest receway.

In the above figure, the box size must be at least 24" in length.

**Code requirement for sizing pull boxes.**

Figure 3.15

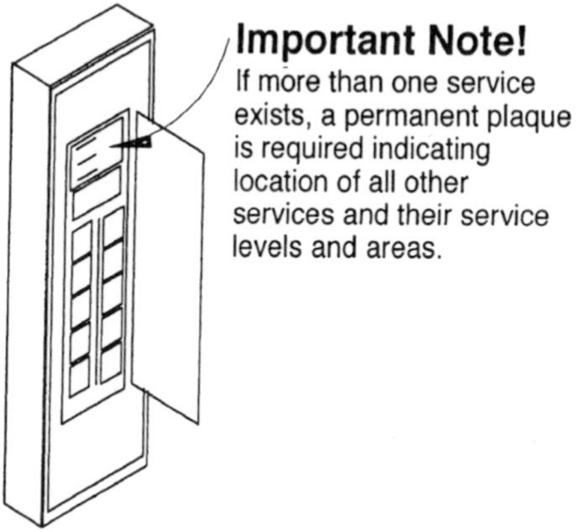

Important Note!
If more than one service exists, a permanent plaque is required indicating location of all other services and their service levels and areas.


**If a building has more than one service distribution panel, the Code requires a permanent plaque. See NEC 230.2.**

## SWITCHING ARRANGEMENTS

Switches must interrupt only the hot (black or red) wire in a circuit. When a circuit must be controlled from two or more locations, three- and four-way switches are used. So that such switches will work properly, they must be properly wired. Hot wires are always attached to the common terminal (darker colored). The lighter colored terminals are for connection of the "traveler" wires. To avoid confusing them with the hot conductor, travelers should be some color other than black, red, green, or white. See illustration following.

Figure 3.16

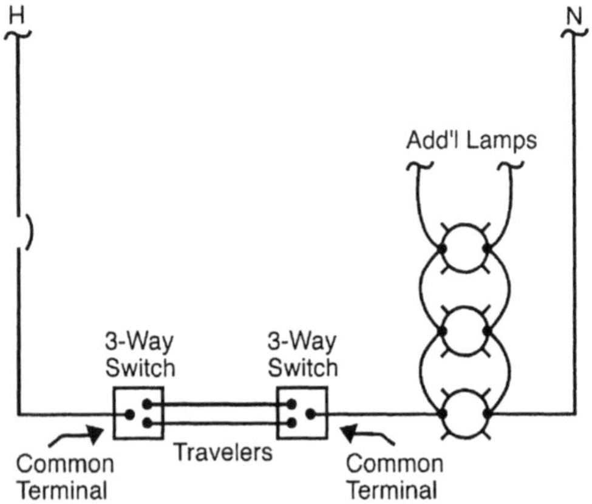


**A three-way switching arrangement. Proper identification of common terminals (usually darker color) is very important. Connect only the "hot" conductors to them. A "traveler" conductor attached to a common terminal will break continuity in certain switch positions.**

Figure 3.17

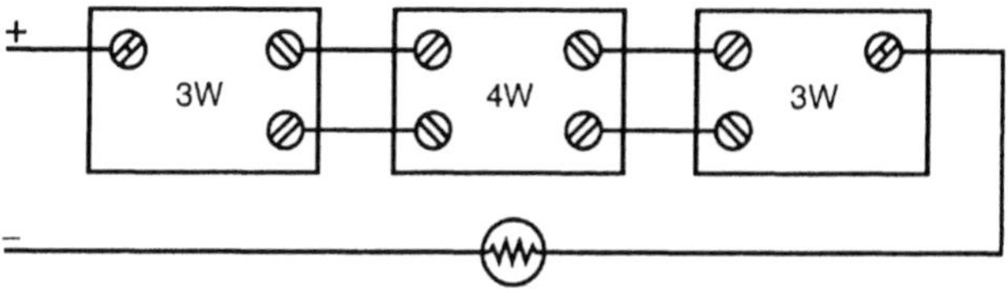


**Wiring up four-way switches are easy since only travelers are connected to them. For additional control points, add four-way switches.**

Figure 3.18

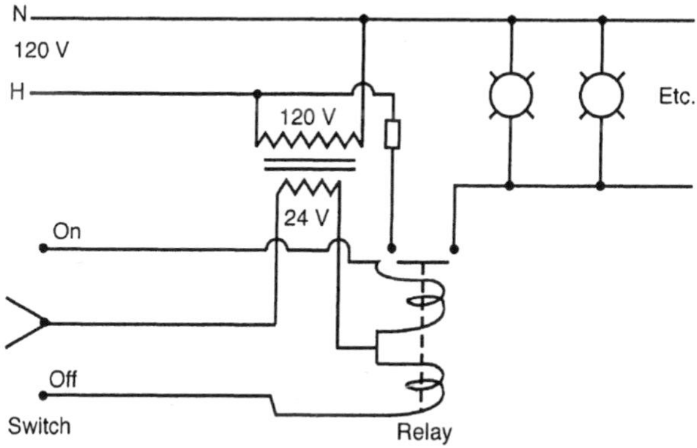


**Schematic of a remote control circuit using low voltage to control a 120 volt lighting circuit.**

Figure 3.19

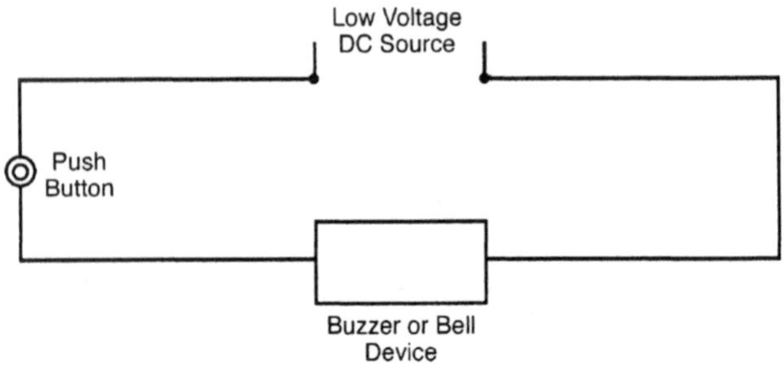


**Simple DC circuit for operating a bell or buzzer.**

Figure 3.20

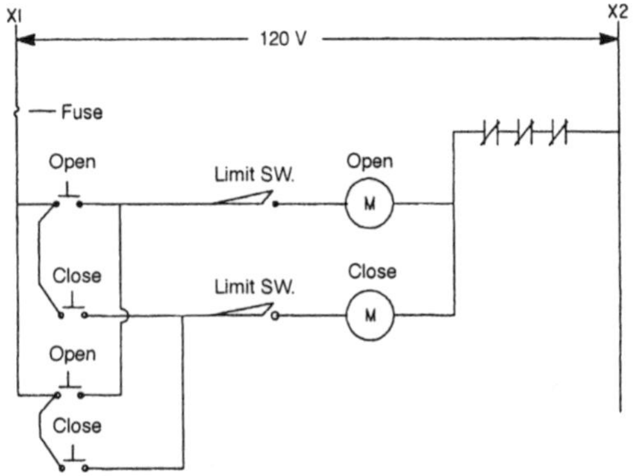


**Schematic for a garage door opener circuit controlled from inside and outside a building.**

## GROUND FAULT CIRCUIT INTERRUPTERS

The Code requires the placement of GFCIs in many locations in buildings (See NEC 210.8):

1. All 125 volt, single-phase, 15 and 20 ampere receptacles in bathrooms, roof tops and outdoors where there is direct access to grade.
2. At least one basement receptacle.
3. All kitchen countertop receptacles.
4. All 125 volt, single-phase, 15 or 20 ampere receptacles in garages unless not readily accessible or dedicated for a cord-and-plug-connected fixed appliance.
5. All 125 volt, single-phase, 15 and 20 ampere receptacles that are not part of permanent wiring.

These requirements can be met by installing GFCI circuit breakers or GFCI receptacles. The following illustrations show the circuitry of both:

Figure 3.21

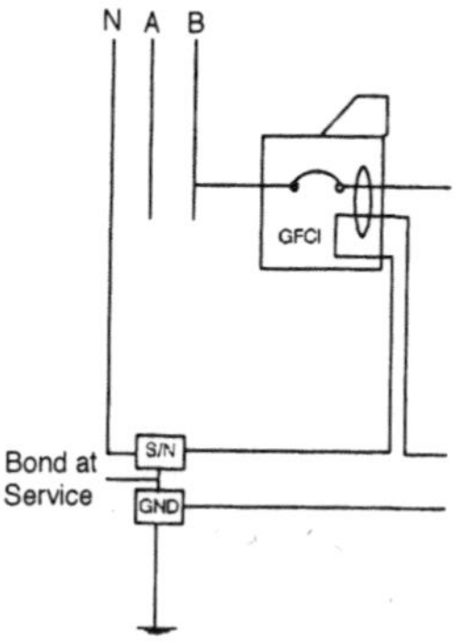


**Circuitry (typical) of a GFCI circuit breaker.**

Figure 3.22

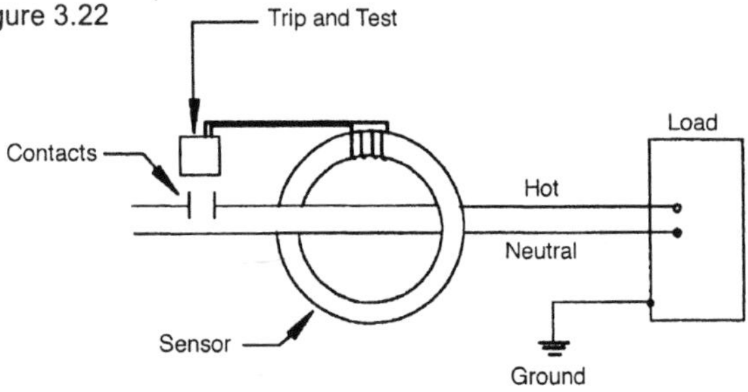


**Typical circuitry of GFCI installed in place of a receptacle.**

Table 3.1

**Maximum Number of Conductors According to Metal Box Size**

| Box Dimension (Inches) | Capacity (Cu. Inches) | AWG | | | |
|---|---|---|---|---|---|
| | | 14 | 12 | 10 | 8 |
| 3 1/4 x 1 1/2 Octagonal | 10.9 | 5 | 4 | 4 | 3 |
| 3 1/2 x 1 1/2 Octagonal | 11.9 | 5 | 5 | 4 | 3 |
| 4 x 1 1/2 Octagonal | 17.1 | 8 | 7 | 6 | 5 |
| 4 x 2 1/8 Octagonal | 23.6 | 11 | 10 | 9 | 7 |
| 4 x 1 1/2 Square | 22.6 | 11 | 10 | 9 | 7 |
| 4 x 2 1/8 Square | 31.9 | 15 | 14 | 12 | 10 |
| 4 11/16 x 1 1/2 Square | 32.2 | 16 | 14 | 12 | 10 |
| 4 11/16 x 2 1/8 Square | 46.4 | 23 | 20 | 18 | 15 |
| 3 x 2 x 1 1/2 Device | 7.9 | 3 | 3 | 3 | 2 |
| 3 x 2 x 2 Device | 10.7 | 5 | 4 | 4 | 3 |
| 3 x 2 x 2 1/4 Device | 11.3 | 5 | 5 | 4 | 3 |
| 3 x 2 x 2 1/2 Device | 13.0 | 6 | 5 | 5 | 4 |
| 3 x 2 x 2 3/4 Device | 14.6 | 7 | 6 | 5 | 4 |
| 3 x 2 x 3 ½ Device | 18.3 | 9 | 8 | 7 | 6 |
| 4 x 2 1/8 x 1 1/2 Device | 11.1 | 5 | 4 | 4 | 3 |
| 4 x 2 1/8 x 1 7/8 Device | 13.9 | 6 | 6 | 5 | 4 |
| 4 x 2 1/8 x 2 1/8 Device | 15.6 | 7 | 6 | 6 | 5 |

The NEC clearly specifies the number and sizes of conductors that can be housed in various sizes and shapes of boxes. The table on previous page applies only where no fittings or devices such as cable clamps, switches, or receptacles are included in the box. Where included, the number of conductors in the box must be reduced. Additional deductions must be made for the presence of one or more equipment grounding conductors.

Figure 3.23

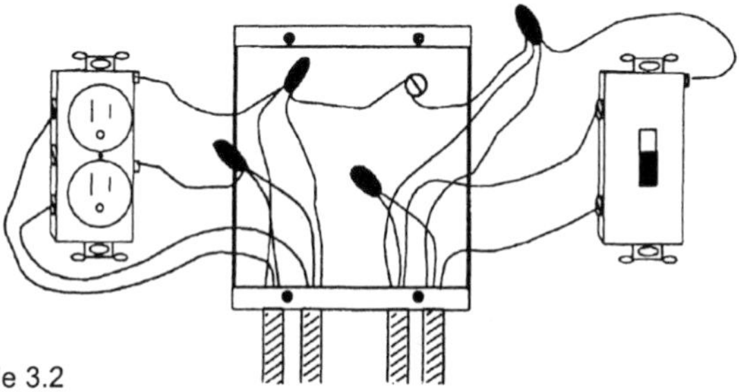

Table 3.2

**How to Determine Minimum Box Size Based on Box Fill**

| Item | Quantity | Count As | Cu. In. Ea. | Total Cu. In. |
|---|---|---|---|---|
| Receptacle Conductors | 4 | 4 | 2.25 | 9.0 |
| Switch Conductors | 4 | 4 | 2.0 | 8.0 |
| Receptacle Yoke | 1 | 2 | 2.25 | 4.5 |
| Switch Yoke | 1 | 2 | 2.0 | 4.0 |
| Equipment Grounds | All | 1 | 2.25 | 2.25 |
| Cable Clamps | 4 | 1 | 2.25 | 2.25 |
| Total → | | | | 30.0 Cu. In. |

Table 3.3

| Conductor Sizes Required for Branch Circuits | | |
|---|---|---|
| **Circuit Rating (Amperes)** | **Minimum AWG Size*** | **Tap Sizes** |
| 15 | 14 | 14 |
| 20 | 12 | 14 |
| 30 | 10 | 14 |
| 40 | 8 | 12 |
| 50 | 6 | 12 |
| *Copper conductors only | | |

Table 3.4

## Overcurrent Protection for Branch Circuits

| Fuse/Circuit Breaker Size (A) | Lampholder Types Allowed | Receptacle Rating (AMP)** |
|---|---|---|
| 15 | Any | 15 Maximum |
| 20 | Any | 15 or 20 |
| 30 | Heavy Duty | 30 |
| 40 | Heavy Duty | 40 or 50 |
| 50 | Heavy Duty | 50 |
| **Refer to NEC Section 410.30(C) for ratings of cord-connected, electric-discharge luminaires (lighting fixtures). | | |

Table 3.5

## Maximum Load Restrictions, Various Branch Circuit Sizes

| Maximum Load in Amperes | NEC Reference |
|---|---|
| 15 | 210.23(A) |
| 20 | 210.23(A) |
| 30 | 210.23(B) |
| 40 | 210.23(C) |
| 50 | 210.23(C) |

Table 3.6

**MINIMUM LIGHTING LOADS**

| Building Type | Volt-Amperes Per Sq. Ft. |
|---|---|
| Auditoriums | 1 |
| Barber/Beauty Shops | 3.5 |
| Churches | 3 |
| Club | 2 |
| Courtroom | 2 |
| Dwelling | 3 |
| Garage | 0.5 |
| Hospital | 2 |
| Hotel/Motel | 2 |
| Industrial/Commercial | 2 |
| Office Buildings | 3.5 |
| Restaurant | 2 |
| School | 3 |
| Store | 3 |
| Warehouse | 0.25 |

Table 3.7

**MINIMUM DISTANCES BETWEEN SUPPORTS FOR CONDUIT & CABLE**

| | |
|---|---|
| Liquid-tight Flexible Conduit | 4.5 ft., 12 in. from outlet box |
| Wireways | 15 ft. vertically; 5 ft. horizontally |
| Nonmetallic wireways | 4 ft. vertical; 3 ft. horizontal |
| Busways | 5 ft. |
| Cableways | 12 ft. |
| Flexible Metal Conduit | 4.5 ft., 12 in. from outlet box |
| Electrical Metal Tubing | 10 ft., 3 ft. from outlet box |
| Rigid Nonmetallic Conduit | |
| 1/2 to 1 in. | 3 ft. |
| 1 3/4 to 2 in. | 5 ft. |
| 2 1/2 to 3 in. | 6 ft. |
| 3 1/2 to 5 in. | 7 ft. |
| 6 in. | 8 ft. |
| Rigid Metal Conduit | |
| 1/2 to 3/4 in. | 10 ft. and within 3 ft. of outlet |
| 1 in. | 12 ft. " " " " " |
| 1 1/4 to 1 1/2 in. | 14 ft. " " " " " |
| 2 to 2 1/2 in. | 16 ft. " " " " " |
| 3 in. and up | 20 ft. " " " " " |
| Intermediate Metal Conduit | 10 ft. and within 3 ft. of outlet |
| Metal-clad Cable | 6 ft. & within 12 in. of outlet box |
| Nonmetallic Sheathed Cable | 4.5 ft. & 12 in. from outlet box |
| Armored Cable | 4.5 ft. & 12 in. from outlet box |
| Mineral-insulated, Metal-sheathed Cable | 6 ft. |
| Electrical Nonmetallic Tubing | 3 ft. |
| Cablebus | 3 ft. horizontal & 1.5 ft. vertical |

Table 3.8

| MINIMAL SIZING OF EQUIPMENT GROUNDING CONDUCTORS FOR RACEWAY AND EQUIPMENT GROUNDING | | |
|---|---|---|
| **Rating/Setting of Automatic Overcurrent Devices (Not Exceeding Listed AMPS)** | **Wire Size in Copper** | **Wire Size Aluminum or Copper-Clad Aluminum** |
| 15 | 14 | 12 |
| 20 | 12 | 10 |
| 30 | 10 | 8 |
| 40 | 10 | 8 |
| 60 | 10 | 8 |
| 100 | 8 | 6 |
| 200 | 6 | 4 |
| 300 | 4 | 2 |
| 400 | 3 | 1 |
| 500 | 2 | 1/0 |
| 600 | 1 | 2/0 |
| 800 | 1/0 | 3/0 |
| 1000 | 2/0 | 4/0 |
| 1200 | 3/0 | 250 kcmil |
| 1600 | 4/0 | 350 kcmil |
| 2000 | 250 kcmil | 400 kcmil |
| 2500 | 350 kcmil | 600 kcmil |
| 3000 | 400 kcmil | 600 kcmil |
| 4000 | 500 kcmil | 800 kcmil |
| 5000 | 700 kcmil | 1200 kcmil |
| 6000 | 800 kcmil | 1200 kcmil |

Table 3.9

| Ampacities of Insulated Copper Conductors Rated at 0-2000 V, 60° to 90°C (140° to 194°F)[1] | | | | |
|---|---|---|---|---|
| **Size** | **140°F (60°C)** | **167°F (75°C)** | **185°F (85°C)** | **194°F (90°)** |
| **AWG KCMIL** | **TYPE** TW[2] UF[2] | **TYPE** FEPW[2] RH[2] RHW[2] THHW[2] THW[2] THWN[2] XHHW[2] USE[2] ZW[2] | **TYPE** V | **TYPE** TA TBS SA SIS FEP[2] FEPB[2] RHH[2] THHN[2] THHW[2] XHHW[2] |
| 18 | --- | --- | --- | 14 |
| 16 | --- | --- | 18 | 18 |
| 14 | 20[2] | 20[2] | 25 | 25[2] |
| 12 | 25[2] | 25[2] | 30 | 30[2] |
| 10 | 30 | 35[2] | 40 | 40[2] |
| 8 | 40 | 50 | 55 | 55 |
| 6 | 55 | 65 | 70 | 75 |
| 4 | 70 | 85 | 95 | 95 |
| 3 | 85 | 100 | 110 | 110 |
| 2 | 95 | 115 | 125 | 130 |
| 1 | 110 | 130 | 145 | 150 |
| 1/0 | 125 | 150 | 165 | 170 |
| 2/0 | 145 | 175 | 190 | 195 |
| 3/0 | 165 | 200 | 215 | 225 |
| 4/0 | 195 | 230 | 250 | 260 |

1) No more than 3 conductors in raceway, cable, or earth (direct burial) at ambient 86°F.
2) Overcurrent protection for this type conductor, unless specifically permitted somewhere in Code, shall not exceed 15 A for #14, 20 A for #12, & 30 A for #10.

Table 3.9 (cont.)

| Ampacities of Insulated Copper Conductors Rated at 0-2000 V, 60° to 90°C (140° to 194°F)[1] (cont.) | | | | |
|---|---|---|---|---|
| Size | 140°F (60°C) | 167°F (75°C) | 185°F (85°C) | 194°F (90°) |
| AWG KCMIL | TYPE TW[2] UF[2] | TYPE FEPW[2] RH[2] RHW[2] THHW[2] THW[2] THWN[2] XHHW[2] USE[2] ZW[2] | TYPE V | TYPE TA TBS SA SIS FEP[2] FEPB[2] RHH[2] THHN[2] THHW[2] XHHW[2] |
| 250 | 215 | 255 | 275 | 290 |
| 300 | 240 | 285 | 310 | 320 |
| 350 | 260 | 310 | 340 | 350 |
| 400 | 280 | 335 | 365 | 380 |
| 500 | 320 | 380 | 415 | 430 |
| 600 | 355 | 420 | 460 | 475 |
| 700 | 385 | 460 | 500 | 520 |
| 750 | 400 | 475 | 515 | 535 |
| 800 | 410 | 490 | 535 | 555 |
| 900 | 435 | 520 | 565 | 585 |
| 1000 | 455 | 545 | 590 | 615 |
| 1250 | 495 | 590 | 640 | 665 |
| 1500 | 520 | 625 | 680 | 705 |
| 1750 | 545 | 650 | 705 | 735 |
| 2000 | 560 | 665 | 725 | 750 |

1) No more than 3 conductors in raceway, cable, or earth (direct burial) at ambient 86°F.
2) Overcurrent protection for this type conductor, unless specifically permitted somewhere in Code, shall not exceed 15 A for #14, 20 A for #12, & 30 A for #10.

Table 3.9 (cont.)

| Ampacities of Insulated Copper Conductors Rated at 0-2000 V (cont.) | | | | |
|---|---|---|---|---|
| **Ampacity Correction Factors** | | | | |
| For temps. other than 86°F multiply ampacities by factors below. | | | | |
| **Temperature (°F)** | **TYPE** TW[2] UF[2] | **TYPE** FEPW[2] RH[2] RHW[2] THHW[2] THW[2] THWN[2] XHHW[2] USE[2] ZW[2] | **TYPE** V | **TYPE** TA TBS SA SIS FEP[2] FEPB[2] RHH[2] THHN[2] THHW[2] XHHW[2] |
| 70-77 | 1.08 | 1.05 | 1.04 | 1.04 |
| 79-86 | 1.00 | 1.00 | 1.00 | 1.00 |
| 88-95 | .91 | .94 | .95 | .96 |
| 97-104 | .82 | .88 | .90 | .91 |
| 106-113 | .71 | .82 | .85 | .87 |
| 115-122 | .58 | .75 | .80 | .82 |
| 124-131 | .41 | .67 | .74 | .76 |
| 133-140 | --- | .58 | .67 | .71 |
| 142-158 | --- | .33 | .52 | .58 |
| 160-176 | --- | --- | .30 | .41 |

Table 3.10

| **Ampacities of Insulated Aluminum or Copper-clad Aluminum Rated at 0-2000 V at 140° to 194°F No more than 3 Conductors in Raceway, Cable, or Earth (Direct Burial) at Ambient Temperature of 86°F (30°C)** | | | | |
|---|---|---|---|---|
| **Size** | **Temperature Rating of Conductor** | | | |
| **AWG kcmil** | **140°F (60°C)** | **167°F (75°C)** | **185°F (85°C)** | **194°F (90°C)** |
| | **TYPES** TW[1] UF[1] | **TYPES** RH[1] RHW[1] THHW[1] THW[1] THWN[1] XHHW[1] USE[1] | **TYPE** V | **TYPES** TA TBS SA SIS RHH[1] THHW[1] THHN[1] XHHW[1] |
| 12 | 20[1] | 20[1] | 25 | 25[1] |
| 10 | 25 | 30[1] | 30 | 35[1] |
| 8 | 30 | 40 | 40 | 45 |
| 6 | 40 | 50 | 55 | 60 |
| 4 | 55 | 65 | 75 | 75 |
| 3 | 65 | 75 | 85 | 85 |
| 2 | 75 | 90 | 100 | 100 |
| 1 | 85 | 100 | 110 | 115 |
| 1/0 | 100 | 120 | 130 | 135 |
| 2/0 | 115 | 135 | 145 | 150 |
| 3/0 | 130 | 155 | 170 | 175 |
| 4/0 | 150 | 180 | 195 | 205 |
| 250 | 170 | 205 | 220 | 230 |
| 300 | 190 | 230 | 250 | 255 |
| 350 | 210 | 250 | 270 | 280 |

1) Unless permitted elsewhere in Code, conductor types marked (1) shall not exceed 15 A for #12 and 25 A for #10 after any correction factors for ambient temperature or number of conductors have been applied.

Table 3.10 (cont.)

| Ampacities of Insulated Aluminum or Copper-clad Aluminum Rated at 0-2000 V at 140° to 194°F No more than 3 Conductors in Raceway, Cable, or Earth (Direct Burial) at Ambient Temperature of 86°F (30°C) (cont.) | | | | |
|---|---|---|---|---|
| **Size** | **Temperature Rating of Conductor** | | | |
| **AWG kcmil** | **140°F (60°C)** | **167°F (75°C)** | **185°F (85°C)** | **194°F (90°C)** |
| | **TYPES** TW[1] UF[1] | **TYPES** RH[1] RHW[1] THHW[1] THW[1] THWN[1] XHHW[1] USE[1] | **TYPE** V | **TYPES** TA TBS SA SIS RHH[1] THHW[1] THHN[1] XHHW[1] |
| 400 | 225 | 270 | 295 | 305 |
| 500 | 260 | 310 | 335 | 350 |
| 600 | 285 | 340 | 370 | 385 |
| 700 | 310 | 375 | 405 | 420 |
| 750 | 320 | 385 | 420 | 435 |
| 800 | 330 | 395 | 430 | 450 |
| 900 | 355 | 425 | 465 | 480 |
| 1000 | 375 | 445 | 485 | 500 |
| 1250 | 405 | 485 | 525 | 545 |
| 1500 | 435 | 520 | 565 | 585 |
| 1750 | 455 | 545 | 595 | 615 |
| 2000 | 470 | 560 | 610 | 630 |

1) Unless permitted elsewhere in Code, conductor types marked (1) shall not exceed 15 A for #12 and 25 A for #10 after any correction factors for ambient temperature or number of conductors have been applied.

Table 3.10 (cont.)

| Ampacities of Insulated Aluminum or Copper-clad Aluminum Rated at 0-2000 V (cont.) | | | | |
|---|---|---|---|---|
| **Ampacity Correction Factors** | | | | |
| For temps. other than 86°F multiply ampacities by factors below. | | | | |
| **Temp. °F** | **TYPES** TW[1] UF[1] | **TYPES** RH[1] RHW[1] THHW[1] THW[1] THWN[1] XHHW[1] USE[1] | **TYPE** V | **TYPES** TA TBS SA SIS RHH[1] THHW[1] THHN[1] XHHW[1] |
| 70-77 | 1.08 | 1.05 | 1.04 | 1.04 |
| 79-86 | 1.00 | 1.00 | 1.00 | 1.00 |
| 88-95 | .91 | .94 | .95 | .96 |
| 97-104 | .82 | .88 | .90 | .91 |
| 106-113 | .71 | .82 | .85 | .87 |
| 115-122 | .58 | .75 | .80 | .82 |
| 124-131 | .41 | .67 | .74 | .76 |
| 133-140 | --- | .58 | .67 | .71 |
| 142-158 | --- | .33 | .52 | .58 |
| 160-176 | --- | --- | .30 | .41 |

Table 3.11

| Allowable Current in Flexible Cords & Cables** | | | | |
|---|---|---|---|---|
| **Size AWG** | **Thermo-set Type TS**<br><br>**Thermo-plastic Types TPT TST** | **Thermoset Types**<br>**C E EO PD S SJ SJO SJOO SO SOO SP-1 SP-2 SP-3 SRD SV SVO SVOO**<br>**Thermoplastics**<br>**ET ETLB ETP ETT SE SEO SJE SJEO SJT SJTO SJTOO SPE-1 SPE-2 SPE-3 SPT-1 SPT-2 SPT-3 ST SRDE SRDT STO STOO SVE SVEO SVT SVTO SVTOO** | | **Types AFS AFSJ HPD HPN HS HSJ HSJO HSJOO HSO HSOO** |
| | | A[2] | B[2] | |
| 27[1] | 0.5 | --- | --- | --- |
| 20 | --- | 5[3] | ---[4] | --- |
| 18 | --- | 7 | 10 | 10 |
| 17 | --- | --- | 12 | --- |
| 16 | --- | 10 | 13 | 15 |
| 15 | --- | --- | --- | 17 |
| 14 | --- | 15 | 18 | 20 |
| 12 | --- | 20 | 25 | 30 |
| 10 | --- | 25 | 30 | 35 |
| 8 | --- | 35 | 40 | --- |
| 6 | --- | 45 | 55 | --- |
| 4 | --- | 60 | 70 | --- |
| 2 | --- | 80 | 95 | --- |

1) Tinsel cord.
2) Currents allowed under "A" apply to 3-wire cords and other multi-conductor cords connected to utilization equipment so that only 3-wire cords are carrying current. Allowable currents under "B" apply to 2-wire cords and other multi-conductor cords connected to utilization equipment so that only 2 conductors are current carrying.
3) Elevator cables only.
4) 7 amperes for elevator cables only; 2 A for other types.

**See NEC Article 400.

## UNIT 4

# ELECTRIC MOTORS AND MOTOR CONTROLS

## ELECTRIC MOTOR TERMS

***AIR-OVER***: Motors designed for fan or blower service. Must be located in the airstream for proper motor cooling.

***AMBIENT:*** Temperature of airspace around a motor. Most motors are designed to operate where temperatures are not over 104°F (40°C).

***BEARINGS:*** Devices for reducing friction between stationary and rotating parts of a motor. Types include:

1. *SLEEVE: Preferred where low noise level is important. Usually can be installed in any position.*
2. *BALL:* Used where higher load capacity is needed or periodic lubrication is impractical. Rings and shields are used to keep out dirt.
3. *UNIT:* A long, single-sleeve bearing. Suitable for fan duty only. Usually suited for mounting in all positions.

***EFFICIENCY:*** Ratio of output power to input power. Usually expressed as a percentage. Measures how well the motor converts the electrical energy to mechanical energy at the output shaft.

***ENCLOSURE:*** The motor housing. Types include:

1. *DRIPPROOF (DP):* Ventilation openings in end shields and shell are placed so liquids falling within a 15-degree angle from vertical will not affect performance.
2. *TOTALLY ENCLOSED (TE):* No vents in housing--but not airtight.

***FRAME:*** Motor mounting dimensions that usually refer to the NEMA standardization system; makes replacement easier.

***FULL-LOAD AMPS (FLA):*** Line current drawn by a motor when operating at its rated load and voltage. Shown on motor nameplate. Important for proper wire size selection and motor starter heater selection.

***HERTZ (Hz):*** Cycles per second of AC power—usually 60 Hz USA and 50 Hz overseas.

***INSULATION:*** Refers to motors, usually classified by maximum allowable total operating temperatures—Class A, 220°F; Class B, 266°F; Class F, 311°F; Class H, 356°F.

***MOTOR SPEEDS:***

1. *SYNCHRONOUS:* Theoretical maximum speed at which an induction-type motor can operate. This speed is determined by the power line frequency and motor design (number of poles) and calculated by the formula

$$Syn.\ RPM = \frac{Frequency\ in\ Hz \times 120}{No.\ of\ Poles}$$

2. *FULL-LOAD:* Nominal speed at which an induction motor operates under rated input and load conditions. Speed will always be less than synchronous speed and will vary according to the rating and characteristics of the motor.

***MOTOR TYPES:*** Classification is determined by operating characteristics and/or type of power required. There are eight in all, including types and sub-types:

1. *INDUCTION MOTORS FOR AC:* Most common type; speed remains relatively constant as load changes. Induction motors include:
   A. *Single Phase* (five types)*:*
      a. *Shaded pole*--low starting torque; cheap; often used in direct drive fans and blowers, and in small gear motors.
      b. *Permanent split capacitor (PSC)*--Similar to shaded pole in application and performance but more efficient; lower line current and higher hp capabilities.
      c. *Split-phase start, induction-run* (or simply split phase)—moderate starting torque, high breakdown torque, medium starting current; used on easy-starting equipment.
      d. *Capacitor-start, induction-run* (or simply capacitor start or capacitor)--high starting and breakdown torque, medium starting current; used on hard-starting applications.
      e. *Capacitor-start, capacitor-run*--performance and applications similar to capacitor-start, induction-run, but having higher efficiency; generally used on higher hp single-phase ratings.
   B. *Three phase:* Operate on three-phase power only; high starting and breakdown torque, high efficiency, medium starting current, simple, rugged design, long life; used in all types of industrial applications.

2. *DIRECT CURRENT (DC):* Usable only with direct current power; usually used on adjustable speed applications.

3. *AC/DC (AC series or Universal):* Operates on AC or DC power; high speed, usually 5000 rpm or higher; brush type; speed drops rapidly as load increases; used on electric drills and saws where high output and small size are desired.

***MOUNTING (Mtg):*** Bracket that secures electric motor to its base. There are five types:

1. *RIGID:* Metal base bolted or welded to motor shell.
2. *RESILIENT (Res):* (Sometimes called rubber or rbr.) Motor shell is isolated from base by vibration-absorbing material, such as rubber rings on the end shields, to reduce vibration.
3. *FACE or FLANGE:* Mounting that attaches to flat mounting surface on the end of the shaft. Holes allow easy, secure mounting to driven equipment. Commonly used on jet pumps, oil burners, and gear reducers.
4. *STUD:* Attaches to motor with bolts extending from front or rear of shell. Often used on small, direct-drive fans and blowers.
5. *YOKE:* Motor shell has tab or ears welded to it. Used on fan-duty motors, allowing bolting to a fan column or bracket.

***POWER:*** The energy used to do work; also the rate of doing work; measurement is in watts, horsepower, etc.

***POWER FACTOR:*** A ratio (real power, in watts, divided by apparent power, in volt-amperes).

***ROTATION:*** Direction that motor shaft turns; CW = clockwise, CCW = counterclockwise; Rev = reversible.

***SERVICE FACTOR (SF):*** Measure of reserve margin built into a motor. Those rated over 1.0 SF have more than normal margin; are used where unusual conditions such as occasional high or low voltage, momentary overloads, etc. are likely to occur.

***SEVERE DUTY:*** A totally enclosed motor with extra protection (shaft slinger, gasketed terminal box, etc.) to keep out contaminants. Used in dirty, wet, or other contaminated environments.

***TEMPERATURE RISE:*** Amount by which a motor operating under rated conditions is hotter than its surroundings. More and more manufacturers are replacing the Rise rating on the nameplate with a list of the ambient temperature rating, insulation class, and service factor.

***THERMAL PROTECTOR:*** Temperature-sensing device built into the motor that disconnects the motor from its power source if the temperature becomes excessive for any reason. Three basic types:

1. *AUTOMATIC-RESET (Auto):* When motor has cooled, protector restores power automatically. Use not recommended where unexpected restarting would be hazardous.
2. *MANUAL-RESET (Man):* External button must be pressed to restore power. Preferred where unexpected restarting would be hazardous (such as on saws, conveyors, and compressors).
3. *IMPEDANCE (Imp):* Motor design that prevents its burnout in less than 15 days under locked rotor (stalled) conditions.

***TORQUE***: Twist or rotating ability as applied to a shaft; measured in foot-pounds, inch-pounds, ounce-feet, or ounce-inches. In a motor two torque values are important:

1. *LOCKED ROTOR TORQUE* or *STARTING TORQUE:* Torque produced at initial start.
2. *BREAKDOWN TORQUE:* The maximum torque a motor will produce while running without an abrupt drop in speed and power.

***VOLTAGE:*** Pressure in an electrical system; force pushing the electric current through the circuit like pressure in a water system.

Figure 4.1

## Motor Schematics Symbols

| | |
|---|---|
|  | Draw Out Connection |
|  | Medium Voltage Air Circuit Breaker |
| 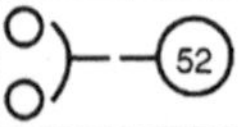 | 600V 3-Phase Air Circuit Breaker |
|  | Medium Voltage Motor Starter |
| 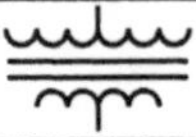 | Transformer Rating As Noted |
|  | Current Transformer Rated As Indicated |
| 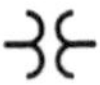 | Potential Transformer Rated As Indicated |
|  | Zero Sequence Current Transformer |
|  | Fuse |
|  | Open or Porcelain Enclosed Fused Cutout |
|  | Oil Filled Fused Cutout |
|  | Capacitor Bank |

Figure 4.1 (cont.)

## Motor Schematics Symbols

| | |
|---|---|
| 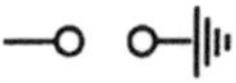 | Lightning Arrestor |
|  | Grounding Resistor |
| 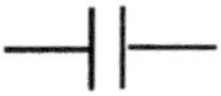 | Electrical Contact - N.O. |
|  | Electrical Contact - N.C. |
|  | Relay Coordination Curve Identification |
|  | Device Located On Mimic Control Panel |
| - L - | Lockout Function |
| - T - | Trip Function |
| - TL - | Trip & Lockout |
| - C - | Close Function |
|  | Ammeter |
|  | Voltmeter |

Figure 4.1 (cont.)

## Motor Schematics Symbols

| | |
|---|---|
| (WHM) | Watt Hour Meter |
| (100 HP) | Induction Motor Horsepower As Indicated |
| [CS] | Control Switch |
| [AS] | Ammeter Switch |
| [VS] | Voltmeter Switch |
| [43] | Manual Transfer or Selector Device |
| (27) | Undervoltage Relay |
| (49) | Thermal Overload Relay |
| (50) | Instantaneous Overcurrent Relay |
| (50G) | Instantaneous Ground Overcurrent Relay |
| (50N) | Instantaneous Neutral Overcurrent Relay |
| (51) | Time Overcurrent Relay |

Figure 4.1 (cont.)

## Motor Schematics Symbols

| | |
|---|---|
| (51G) | Time Overcurrent Ground Relay |
| (51N) | Time Overcurrent Relay on Neutral Circuit |
| (52) | Circuit Breaker |
| (63) | Sudden Pressure Relay |
| (86) | Lockout Relay |
| (87) | Differential Relay |

Figure 4.2

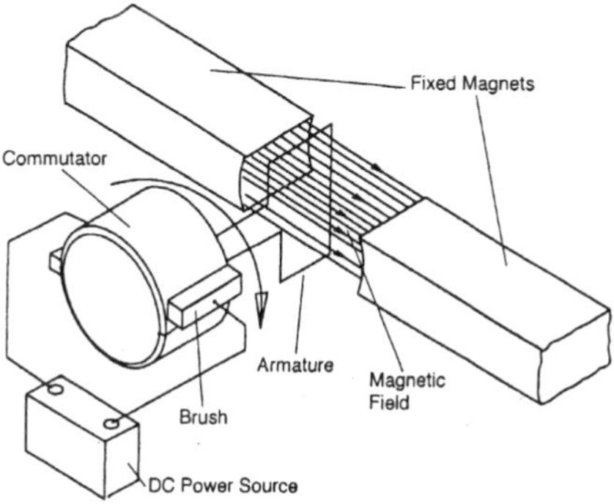


**Parts of a simplified drawing of a DC motor.**

Figure 4.3

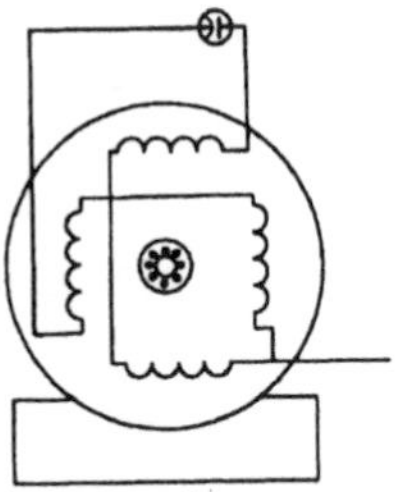

**Split Capacitor Type Motor**

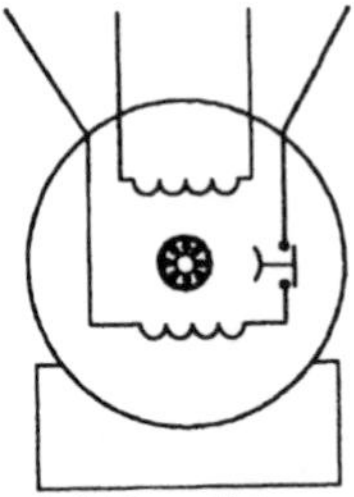

**Split Phase Motor with centrifugal switch**

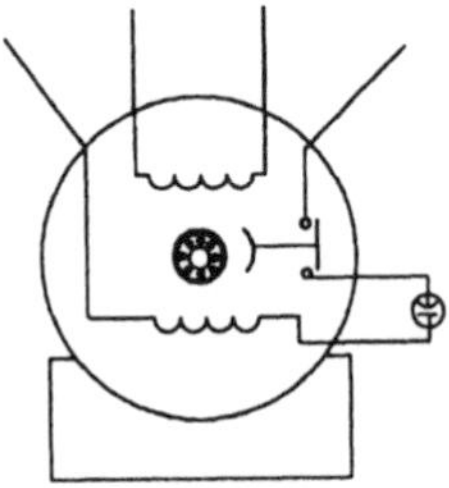

**Capacitor Start Motor with centrifugal switch**

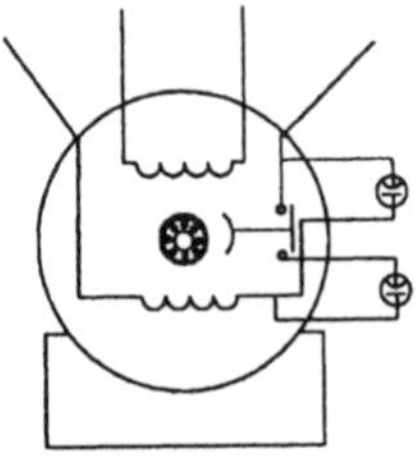

**Capacitor Start and Run Motor**

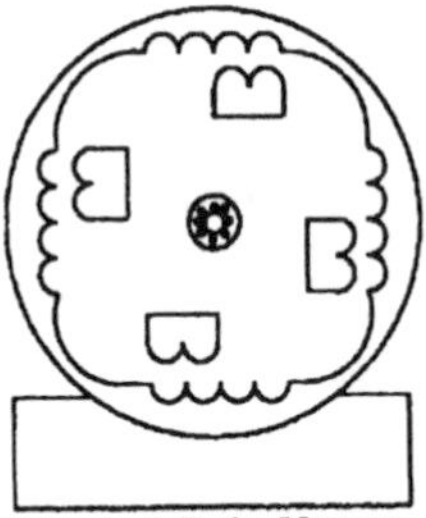

**Shaded-Pole Motor**

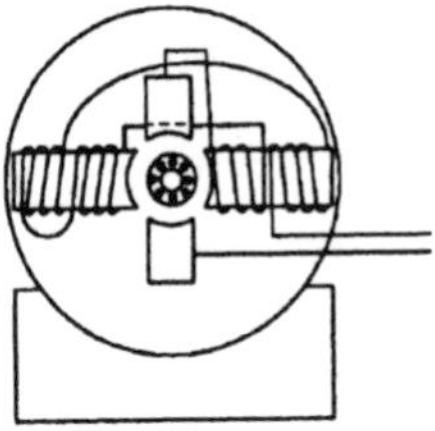

**Universal Motor can operate on either 1ØAC or DC**

Figure 4.3 (cont.)

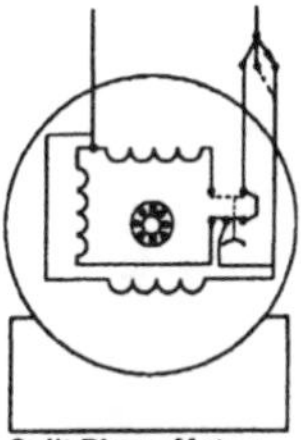
Split Phase Motor with Low-Off-High speed switch and centrifugal switch

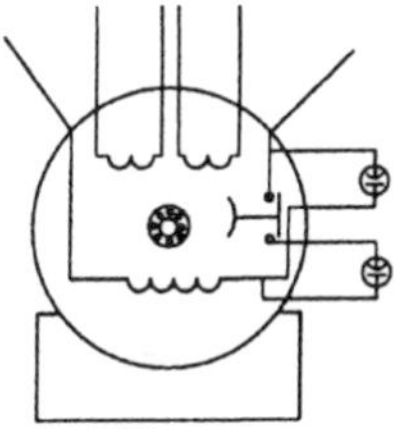
Dual Voltage Capacitor Motor

Fractional HP Shaded-Pole Motor

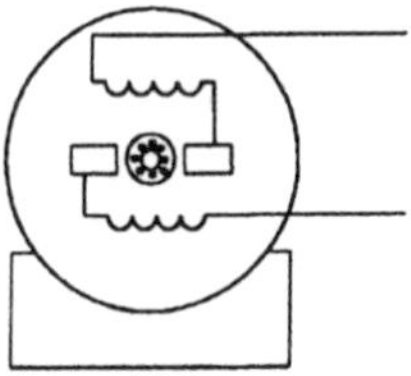
Universal Motor

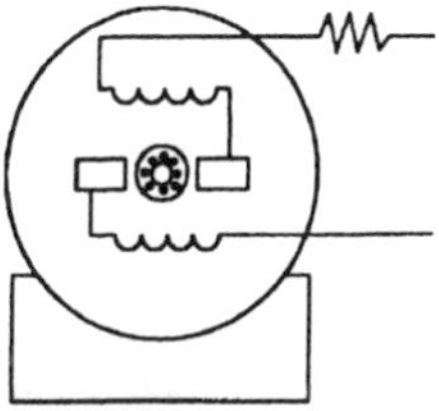
Universal Motor with variable resistor and switch

Figure 4.3 (cont.)

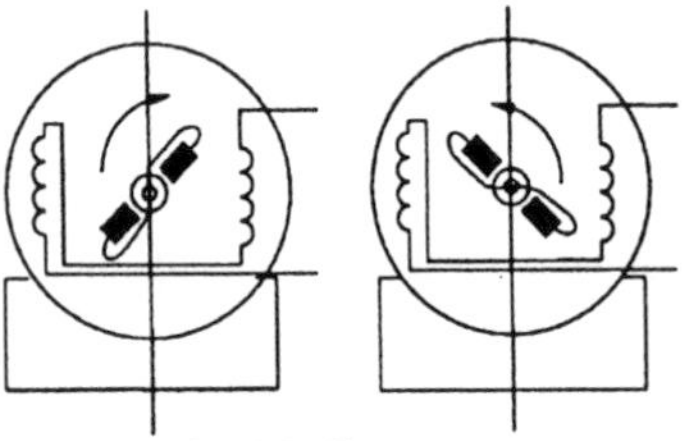

Repulsion Motor
showing rotation of armature
when forward and reversed

**Overview of different types of electric motors (internal and external wiring).**

Figure 4.4

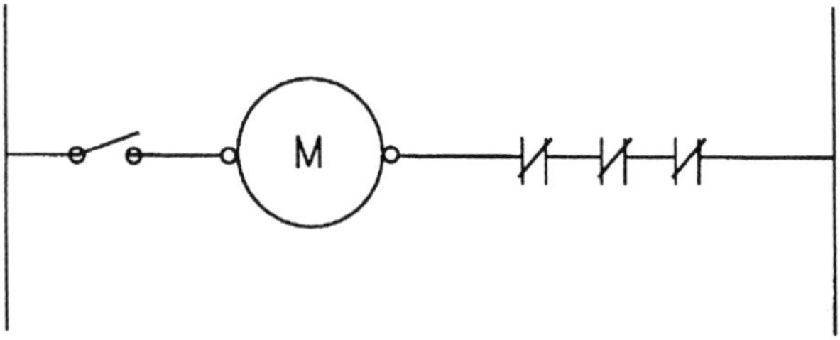


**Typical motor schematic showing (from left) switch, motor, and three normally closed electrical contacts.**

Figure 4.5

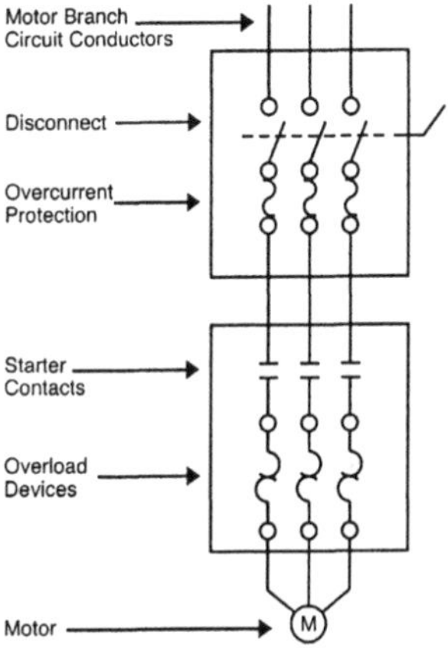


**Schematic of typical motor circuits components.**

## NAMEPLATE CODE & NEMA REQUIREMENTS

Both the National Electrical Code (NEC) and the National Electrical Manufactures Association (NEMA) require that all electric motors have nameplates with the following information: Manufacturer's name, motor type, horsepower rating, frame number, time rating, rpm rating, design letter, frequency (in Hertz), number of phases, insulation class, load rating in amperes, locked rotor code letter, voltage, duty, ambient temperature, and service factor. Often other information, such as serial number, model number, bearing types, and efficiency rating are included. This data should be carefully reviewed before setting and wiring a motor.

Table 4.1a

**Nameplate Data Definitions**

| | |
|---|---|
| **Frame & Type** | NEMA designation for frame and type. |
| **Horsepower** | The power rating of the motor. |
| **Motor Code** | A letter designating starting current required. The higher the locked-rotor kVA, the higher the starting current surge. |
| **Hertz** | Frequency at which an AC motor is designed to operate. |
| **Phase** | The number of phases on which the motor operates. |
| **Speed in RPM** | Revolutions per minute of the motor at full load. |
| **Voltage** | Voltage or voltages at which motor operates. |
| **Thermal Protection** | Indication of protection against overheating, if any. |
| **Amperes (AMPs)** | Rated current at full load. |
| **Duty** | Service rating: continuous or as a specified period of time the motor can be operated. |
| **Ambient Temperature (or Temp. Rise)** | Maximum temperature at which motor should be operated; or permissible temperature rise of the motor above ambient temperature at full load. |
| **Service Factor (SF)** | Amount of overload motor can tolerate on a continuous basis at rated voltage and frequency. |
| **Insulation Class** | Designation of the insulation system used, mostly for convenience in rewinding. |
| **NEMA Design** | Letter that designates (for integral horsepower motors) the motor's characteristics. |

Table 4.1b

| Manufacturer's Name | | | | | |
|---|---|---|---|---|---|
| Model #: | | Serial #: | | Frame/Type: | |
| Hp: | Hz: | V: | Phase: | AMPS: | |
| RPM: | | Code Letter: | | NEMA Design: | |
| Service Factor: | | Amb Temp: | | Insulation Class: | |
| Time Rating: | | Duty: | | Bearing Types: | |
| Thermal Prot: | | Impedance Prot: | | Eff Rating: | |

## MOTOR SELECTION CONSIDERATIONS

Because of the variety of motors available certain factors are important in choosing the correct motor for the job.

### Power Supply

Most power sources are single phase. Single-phase motors of correct voltage will operate on a three-phase system when properly connected. However, three-phase motors cannot be connected to single-phase power. A three-phase motor is less expensive and more durable than the same size single-phase motor, but the extra cost of installing three-phase lines is a consideration.

## Motor Type

A split-phase motor is economical and a good choice where starting loads are light, as in belt-driven fans and blowers, or where loads are not applied until operating speed is reached (table saws and drill presses).

Shaded-pole and permanent split-capacitor motors are usually designed for direct-drive fans and blowers.

Capacitor-start motors are designed for conveyors, air compressors and other devices where heavy loads must be started.

Three-phase motors are used for a wide range of high starting torque applications where three-phase power is available.

## Motor HP and Speed

Check nameplate data when replacing a motor; new motor should have the same specifications. On a new installation, check with the manufacturer or dealer for the appropriate motor. In general, both motor price and size increase as rated HP increases and RPM increases.

## Bearings

Sleeve bearings are designed for loads that are moderate. The sleeve-bearing motor can be mounted in any position. They are quieter and less expensive than ball-bearing motors. The latter is recommended for powering devices that create heavy loads or in locations that are damp and dirty.

## Voltage

Motor rating must match voltage and frequency (Hertz) of supplied power.

## Enclosure

Open, drip-proof motors are designed for use in areas that are fairly dry, clean, and well ventilated. When installed outside, motors should be protected with a cover that does not restrict airflow. Wet or dirty conditions will need totally enclosed, fan-cooled construction. Explosive conditions, no matter how slight, require an explosion-proof motor.

## Shaft and Mounting Dimensions

Select a NEMA frame size or a non-NEMA motor with dimensions matching the old motor or that fits a new application.

Table 4.2

| 110-120 V AC Motor Full-load Running Currents and Recommended Transformer* Ratings | | | | |
|---|---|---|---|---|
| | Single-Phase | | Three-Phase | |
| Hp | AMPS | kVA | AMPS | kVA |
| 1/2 | 9.8 | 1.5 | 4.0 | 3 |
| 3/4 | 13.8 | 2.0 | 5.6 | 3 |
| 1 | 16.0 | 3.0 | 7.2 | 3 |
| 1 1/2 | 20.0 | 3.0 | 10.4 | 3 |
| 2 | 24.0 | 5.0 | 13.6 | 6 |
| 3 | 34.0 | 5.0 | 19.2 | 6 |
| 5 | 56.0 | 7.5 | 30.4 | 9 |
| 7 1/2 | 80.0 | 15 | 44.0 | 15 |
| 10 | 100.0 | 15 | 56.0 | 15 |
| 15 | 135.0 | 25 | 84.0 | 30 |
| 20 | --- | --- | 108.0 | 30 |
| 25 | --- | --- | 136.0 | 45 |
| 30 | --- | --- | 160.0 | 45 |
| 40 | --- | --- | 208.0 | 75 |
| 50 | --- | --- | 260.0 | 75 |
| 60 | --- | --- | --- | --- |
| 75 | --- | --- | --- | --- |
| 100 | --- | --- | --- | --- |

*Recommended kVA rating shown includes a minimum of 10% spare capacity for frequent motor starting.

Table 4.3

| 220-240 V* AC Motor Full-load Running Currents and Recommended Transformer** Ratings | | | | |
|---|---|---|---|---|
| | Single-Phase | | Three-Phase | |
| Hp | AMPS | kVA | AMPS | kVA |
| 1/2 | 4.9 | 1.5 | 2.0 | 3 |
| 3/4 | 6.9 | 2.0 | 2.8 | 3 |
| 1 | 8.0 | 3.0 | 3.6 | 3 |
| 1 1/2 | 10.0 | 3.0 | 5.2 | 3 |
| 2 | 12.0 | 5.0 | 6.8 | 6 |
| 3 | 17.0 | 5.0 | 9.6 | 6 |
| 5 | 28.0 | 7.5 | 15.2 | 9 |
| 7 1/2 | 40.0 | 15 | 22.0 | 15 |
| 10 | 50.0 | 15 | 28.0 | 15 |
| 15 | 68.0 | 25 | 42.0 | 30 |
| 20 | 88.0 | 25 | 54.0 | 30 |
| 25 | 110.0 | 37.5 | 68.0 | 45 |
| 30 | 136.0 | 37.5 | 80.0 | 45 |
| 40 | 176.0 | 50 | 104.0 | 75 |
| 50 | 216.0 | 75 | 130.0 | 75 |
| 60 | --- | --- | 154.0 | 75 |
| 75 | --- | --- | 192.0 | 112.5 |
| 100 | --- | --- | 248.0 | 112.5 |

*Recommended kVA rating shown includes a minimum of 10% spare capacity for frequent motor starting.
**For full-load currents for 200 & 208 V motors, increase corresponding 200-240 V ratings by 15 and 10%.

Table 4.4

| 440-480 V AC Motor Full-load Running Currents and Recommended Transformer* Ratings | | | | |
|---|---|---|---|---|
| | Single-Phase | | Three-Phase | |
| Hp | AMPS | kVA | AMPS | kVA |
| 1/2 | 2.5 | 1.5 | 1.0 | 3 |
| 3/4 | 3.5 | 2.0 | 1.4 | 3 |
| 1 | 4.0 | 3.0 | 1.8 | 3 |
| 1 1/2 | 5.0 | 3.0 | 2.6 | 3 |
| 2 | 6.0 | 5.0 | 3.4 | 6 |
| 3 | 8.5 | 5.0 | 4.8 | 6 |
| 5 | 14.0 | 7.5 | 7.6 | 9 |
| 7 1/2 | 21.0 | 15 | 11.0 | 15 |
| 10 | 26.0 | 15 | 14.0 | 15 |
| 15 | 34.0 | 25 | 21.0 | 30 |
| 20 | 44.0 | 25 | 27.0 | 30 |
| 25 | 55.0 | 37.5 | 34.0 | 45 |
| 30 | 68.0 | 37.5 | 40.0 | 45 |
| 40 | 88.0 | 50 | 52.0 | 75 |
| 50 | 108.0 | 75 | 65.0 | 75 |
| 60 | --- | --- | 77.0 | 75 |
| 75 | --- | --- | 96.0 | 112.5 |
| 100 | --- | --- | 124.0 | 112.5 |

*Recommended kVA rating shown includes a minimum of 10% spare capacity for frequent motor starting.

Table 4.5

| 550-600 V AC Motor Full-load Running Currents and Recommended Transformer* Ratings | | | | |
|---|---|---|---|---|
| | Single-Phase | | Three-Phase | |
| **Hp** | **AMPS** | **kVA** | **AMPS** | **kVA** |
| 1/2 | 2.0 | 1.5 | 0.8 | 3 |
| 3/4 | 2.8 | 2.0 | 1.1 | 3 |
| 1 | 3.2 | 3.0 | 1.4 | 3 |
| 1 1/2 | 4.0 | 3.0 | 2.1 | 3 |
| 2 | 4.8 | 5.0 | 2.7 | 6 |
| 3 | 6.8 | 5.0 | 3.9 | 6 |
| 5 | 11.2 | 7.5 | 6.1 | 9 |
| 7 1/2 | 16.0 | 15 | 9.0 | 15 |
| 10 | 20.0 | 15 | 11.0 | 15 |
| 15 | 27.0 | 25 | 17.0 | 30 |
| 20 | 35.0 | 25 | 22.0 | 30 |
| 25 | 44.0 | 37.5 | 27.0 | 45 |
| 30 | 54.0 | 37.5 | 32.0 | 45 |
| 40 | 70.0 | 50 | 41.0 | 75 |
| 50 | 86.0 | 75 | 52.0 | 75 |
| 60 | --- | --- | 62.0 | 75 |
| 75 | --- | --- | 77.0 | 112.5 |
| 100 | --- | --- | 99.0 | 112.5 |

*Recommended kVA rating shown includes a minimum of 10% spare capacity for frequent motor starting.

Table 4.6

| Full-load Current, One-phase, 60 Hz AC Induction Motors* (In Amperes) | | |
|---|---|---|
| **Horsepower** | **115 Volt** | **230 Volt** |
| 1/6 | 4.4 | 2.2 |
| 1/4 | 5.8 | 2.9 |
| 1/3 | 7.2 | 3.6 |
| 1/2 | 9.8 | 4.9 |
| 3/4 | 13.8 | 6.9 |
| 1 | 16 | 8 |
| 1 1/2 | 20 | 10 |
| 2 | 24 | 12 |
| 3 | 34 | 17 |
| 5 | 56 | 28 |
| 7 1/2 | 80 | 40 |
| 10 | 100 | 50 |

* Above values are for motors running at usual speeds and those with normal torque characteristics. Motors built for very low speeds or high torque may draw higher full-load current. Multi-speed motors will have full-load current varying with speed. In such cases, nameplate current ratings shall be used.
For full-load currents of 208- or 200-V motors, increase 230-V full-load currents 10 and 15 % respectively.
Listed voltages are rated motor voltages. Currents listed shall be permitted for system voltage ranges of 110 to 120 and 220 to 240.

Table 4.7

| Full-load Current in 3-Phase 60 Hz AC Induction Motors* (1/4 - 1 1/2 Horsepower) | | | | | |
|---|---|---|---|---|---|
| | | Full-load Current | | | |
| **HP** | **RPM** | **200 V** | **230 V** | **460 V** | **575 V** |
| 1/4 | 1800 | 1.10 | 0.96 | 0.48 | 0.38 |
| | 1200 | 1.33 | 1.16 | 0.58 | 0.46 |
| | 900 | 1.67 | 1.45 | 0.73 | 0.58 |
| 1/3 | 1800 | 1.33 | 1.16 | 0.58 | 0.47 |
| | 1200 | 1.64 | 1.43 | 0.72 | 0.58 |
| | 900 | 2.01 | 1.75 | 0.88 | 0.71 |
| 1/2 | 1800 | 1.93 | 1.68 | 0.84 | 0.67 |
| | 1200 | 2.38 | 2.07 | 1.04 | 0.83 |
| | 900 | 3.34 | 2.90 | 1.45 | 1.16 |
| 3/4 | 1800 | 2.68 | 2.33 | 1.17 | 0.93 |
| | 1200 | 3.28 | 2.85 | 1.43 | 1.14 |
| | 900 | 3.97 | 3.45 | 1.73 | 1.38 |
| 1 | 3600 | 3.16 | 2.75 | 1.38 | 1.10 |
| | 1800 | 3.51 | 3.05 | 1.53 | 1.22 |
| | 1200 | 4.07 | 3.54 | 1.77 | 1.42 |
| | 900 | 4.30 | 3.74 | 1.87 | 1.50 |
| 1 1/2 | 3600 | 4.80 | 4.17 | 2.09 | 1.67 |
| | 1800 | 4.92 | 4.28 | 2.14 | 1.71 |
| | 1200 | 5.58 | 4.85 | 2.43 | 1.94 |
| | 900 | 6.68 | 5.81 | 2.91 | 2.32 |

*Above are average values to be used, along with like values in NEC, as a guide only in sizing components in the motor branch circuit. Rated full-load current values on a motor nameplate may differ, depending on motor design. ***Always*** use table values in determining overcurrent protection requirements. See NEC 430.6.

Table 4.8

**Full-Load Current for 3-Phase 60 Hz Induction Motors**

**(2 - 15 Horsepower)**

The following "average values" are to be used, along with like values in NEC, as a guide only in determining components of the branch circuit. Rated full-load current values on a motor nameplate may differ, depending on motor design. **Always** use table values to select properly rated overcurrent protection devices. See NEC 430.6.

| | | **Full-Load Current** | | | |
|---|---|---|---|---|---|
| **Hp** | **RPM** | **200 V** | **230 V** | **460 V** | **575 V** |
| 2 | 3600 | 6.39 | 5.56 | 2.78 | 2.22 |
| | 1800 | 6.62 | 5.76 | 2.88 | 2.30 |
| | 1200 | 7.30 | 6.35 | 3.18 | 2.54 |
| | 900 | 8.29 | 7.21 | 3.61 | 2.88 |
| 3 | 3600 | 9.05 | 7.87 | 3.94 | 3.14 |
| | 1800 | 9.53 | 8.29 | 4.14 | 3.32 |
| | 1200 | 10.3 | 8.92 | 4.46 | 3.56 |
| | 900 | 11.7 | 10.2 | 5.09 | 4.08 |
| 5 | 3600 | 14.6 | 12.7 | 6.34 | 5.08 |
| | 1800 | 15.2 | 13.2 | 6.60 | 5.28 |
| | 1200 | 16.2 | 14.1 | 7.05 | 5.64 |
| | 900 | 17.9 | 15.6 | 7.80 | 6.24 |
| 7 1/2 | 3600 | 22.1 | 19.2 | 9.60 | 7.68 |
| | 1800 | 22.2 | 19.3 | 9.70 | 7.72 |
| | 1200 | 23.3 | 20.3 | 10.2 | 8.12 |
| | 900 | 27.4 | 23.8 | 11.9 | 9.51 |
| 10 | 3600 | 28.2 | 24.5 | 12.3 | 9.80 |
| | 1800 | 29.0 | 25.2 | 12.6 | 10.1 |
| | 1200 | 30.6 | 26.6 | 13.3 | 10.6 |
| | 900 | 33.2 | 28.9 | 14.5 | 11.6 |
| | 600 | 38.9 | 33.8 | 16.9 | 13.5 |
| 15 | 3600 | 42.2 | 36.7 | 18.4 | 14.7 |
| | 1800 | 43.8 | 38.1 | 19.1 | 15.2 |
| | 1200 | 45.9 | 39.9 | 20.0 | 16.0 |
| | 900 | 48.2 | 41.9 | 21.0 | 16.8 |
| | 600 | 55.5 | 48.3 | 24.2 | 19.3 |

Table 4.9

| Full-Load Current for 3-Phase 60 Hz AC Induction Motors (20 - 60 Horsepower) |
|---|
| Following "average values" are to be used, along with like values in NEC, as a guide only in determining components of the branch circuit. Rated full-load current values on motor nameplate may differ, depending on motor design. Always use table values in selecting properly rated overcurrent protection devices. See NEC 430.6. |

| | | Full-Load Current | | | | | |
|---|---|---|---|---|---|---|---|
| **Hp** | **RPM** | **200V** | **230V** | **460V** | **575V** | **2200V** | **4000V** |
| 20 | 3600 | 56.4 | 49.0 | 24.5 | 19.6 | 5.2 | 2.9 |
| | 1800 | 58.1 | 50.5 | 25.3 | 20.2 | 5.3 | 3.0 |
| | 1200 | 59.5 | 51.7 | 25.9 | 20.6 | 5.4 | 3.1 |
| | 900 | 62.8 | 54.6 | 27.3 | 21.8 | 5.8 | 3.2 |
| | 600 | 70.7 | 61.5 | 30.8 | 24.6 | 6.4 | 3.5 |
| 25 | 3600 | 68.1 | 59.2 | 29.6 | 23.6 | 6.3 | 3.4 |
| | 1800 | 72.1 | 62.7 | 31.3 | 25.0 | 6.5 | 3.6 |
| | 1200 | 74.4 | 64.7 | 32.3 | 25.8 | 6.7 | 3.7 |
| | 900 | 77.5 | 67.4 | 33.7 | 27.0 | 6.9 | 3.8 |
| | 600 | 82.7 | 71.9 | 35.9 | 28.8 | 8.1 | 4.4 |
| 30 | 1800 | 83.7 | 72.8 | 36.3 | 29.2 | 7.8 | 4.3 |
| | 1200 | 88.7 | 77.1 | 38.6 | 30.8 | 8.0 | 4.4 |
| | 900 | 91.3 | 79.4 | 39.7 | 31.8 | 8.2 | 4.5 |
| | 600 | 101.1 | 87.9 | 43.9 | 35.2 | 9.3 | 5.0 |
| 40 | 1800 | 112.7 | 98.0 | 49.0 | 39.2 | 10.0 | 5.6 |
| | 1200 | 113.9 | 99.0 | 49.5 | 39.6 | 10.3 | 5.7 |
| | 900 | 119.6 | 104.0 | 52.0 | 41.6 | 10.6 | 5.8 |
| | 600 | 130.0 | 113.0 | 56.5 | 45.2 | 11.5 | 6.3 |
| 50 | 1800 | 139.2 | 121.0 | 60.5 | 48.4 | 12.3 | 6.8 |
| | 1200 | 140.3 | 122.0 | 61.1 | 48.8 | 12.4 | 6.8 |
| | 900 | 146.1 | 127.0 | 63.5 | 50.8 | 13.1 | 7.2 |
| | 600 | 158.7 | 138.0 | 69.0 | 55.2 | 14.2 | 7.8 |
| 60 | 1800 | 164.5 | 143.0 | 71.5 | 57.2 | 14.6 | 8.0 |
| | 1200 | 170.2 | 148.0 | 74.0 | 59.2 | 14.9 | 8.2 |
| | 900 | 173.7 | 151.0 | 75.5 | 60.4 | 15.4 | 8.5 |
| | 600 | 186.3 | 162.0 | 81.0 | 64.8 | 16.7 | 9.2 |

Table 4.10

| Full-Load Current for 3-Phase 60 Hz AC Induction Motors (75 - 300 Horsepower) | | | | | | | |
|---|---|---|---|---|---|---|---|
| | | Current at Full Load | | | | | |
| | | Volts | | | | | |
| HP | RPM | 200 | 230 | 460 | 575 | 2200 | 4000 |
| 75 | 1800 | 204.7 | 178 | 89.0 | 71.2 | 18.0 | 9.9 |
| | 1200 | 208.2 | 181 | 90.5 | 72.4 | 18.2 | 10.0 |
| | 900 | 215.1 | 187 | 93.5 | 74.8 | 19.0 | 10.5 |
| | 600 | 228.9 | 199 | 99.5 | 79.6 | 21.0 | 11.6 |
| 100 | 1800 | 268.0 | 233 | 116 | 93.2 | 23.6 | 13.0 |
| | 1200 | 274.9 | 239 | 120 | 95.6 | 24.2 | 13.3 |
| | 900 | 281.8 | 245 | 123 | 98.0 | 24.8 | 13.6 |
| | 600 | 295.6 | 257 | 128 | 103 | 26.4 | 14.5 |
| | 450 | 333.5 | 290 | 145 | 116 | 29.8 | 16.4 |
| 125 | 1800 | 332.4 | 289 | 144 | 115 | 29.2 | 16.1 |
| | 1200 | 342.7 | 298 | 149 | 119 | 29.9 | 16.4 |
| | 900 | 350.8 | 305 | 153 | 122 | 30.9 | 17.0 |
| | 720 | 361.1 | 314 | 157 | 126 | 31.3 | 17.2 |
| | 600 | 368.0 | 320 | 160 | 128 | 32.8 | 18.0 |
| | 450 | 403.7 | 351 | 175 | 140 | 36.0 | 19.8 |
| 150 | 1800 | 397.9 | 348 | 173 | 138 | 34.8 | 19.1 |
| | 1200 | 402.5 | 350 | 175 | 140 | 35.5 | 19.5 |
| | 900 | 417.5 | 363 | 182 | 145 | 37.0 | 20.4 |
| | 720 | 432.4 | 376 | 188 | 150 | 37.0 | 20.4 |
| | 600 | 434.7 | 378 | 189 | 151 | 38.8 | 21.3 |
| | 450 | 480.7 | 418 | 209 | 166 | 42.0 | 23.1 |

These are average values for hp-rated motors of several manufacturers at the more common rated voltages and speeds. These values, together with similar values listed in NEC, should be used as a guide only in choosing suitable parts for the motor branch circuit. Rated full-load current on the motor's nameplate may vary from these values. The nameplate rating should ***always*** be used in determining the rating of devices used for motor overload protection.

Table 4.10 (cont.)

| Full-Load Current for 3-Phase 60 Hz AC Induction Motors (75 - 300 Horsepower) | | | | | | | |
|---|---|---|---|---|---|---|---|
| | | Current at Full Load | | | | | |
| | | Volts | | | | | |
| HP | RPM | 200 | 230 | 460 | 575 | 2200 | 4000 |
| 200 | 1800 | 529.0 | 460 | 230 | 184 | 46.7 | 25.7 |
| | 1200 | 535.9 | 466 | 233 | 186 | 47.0 | 25.9 |
| | 900 | 563.5 | 490 | 245 | 196 | 49.4 | 27.2 |
| | 720 | 568.1 | 494 | 247 | 197 | 49.0 | 27.0 |
| | 600 | 572.7 | 498 | 249 | 199 | 50.9 | 28.0 |
| | 450 | 607.2 | 528 | 264 | 211 | 53.7 | 29.5 |
| 250 | 1800 | 657.8 | 572 | 286 | 229 | 57.5 | 31.6 |
| | 1200 | 667.0 | 580 | 290 | 232 | 58.5 | 32.2 |
| | 900 | 694.6 | 604 | 302 | 242 | 61.5 | 33.8 |
| | 720 | 718.8 | 625 | 312 | 250 | 63.3 | 34.9 |
| | 600 | 724.5 | 630 | 315 | 252 | 63.9 | 35.3 |
| | 450 | 738.0 | 642 | 321 | 259 | 65.3 | 35.9 |
| | 350 | 777.4 | 676 | 338 | 270 | 70.0 | 38.5 |
| 300 | 1800 | 787.8 | 685 | 342 | 274 | 69.0 | 38.0 |
| | 1200 | 800.4 | 696 | 348 | 278 | 70.0 | 38.5 |
| | 900 | 830.3 | 722 | 361 | 289 | 73.5 | 40.4 |
| | 600 | 844.0 | 734 | 367 | 294 | 74.5 | 41.1 |
| | 450 | 874.0 | 760 | 380 | 304 | 76.0 | 41.8 |
| | 360 | 954.5 | 830 | 415 | 332 | 82.8 | 45.5 |

These are average values for hp-rated motors of several manufacturers at the more common rated voltages and speeds. These values, together with similar values listed in NEC, should be used as a guide only in choosing suitable parts for the motor branch circuit. Rated full-load current on the motor's nameplate may vary from these values. The nameplate rating should ***always*** be used in determining the rating of devices used for motor overload protection.

Table 4.11

| Full-load Current Values of 4-Wire, 2-Phase AC Motors | | | | | |
|---|---|---|---|---|---|
| Values in this table are for motors running at speeds that are usual for belted motors and those having normal torque characteristics. However, motors designed for especially low speeds or high torque may need more running current; multi-speed motor full-load currents will vary with speed. In such cases, nameplate current rating shall be used. Current in the common conductor of 2-phase, 3-wire systems will be 1.41 times the given value. Listed voltages are rated motor voltages. Currents listed shall be permitted for system voltage ranges of 110-120, 220-240, 440-480, and 550-600. | | | | | |
| **Full-load Current in Amperes for Induction Type Squirrel-Cage and Wound-Rotor Motors** | | | | | |
| **Hp** | **115 V** | **230 V** | **460 V** | **575 V** | **2300 V** |
| 1/2<br>3/4<br>1 | 4<br>4.8<br>6.4 | 2<br>2.4<br>3.2 | 1<br>1.2<br>1.6 | .8<br>1.0<br>1.3 | -- |
| 1 1/2<br>2<br>3 | 9<br>11.8 | 4.5<br>5.9<br>8.3 | 2.3<br>3<br>4.2 | 1.8<br>2.4<br>3.3 | -- |
| 5<br>7 1/2<br>10 | -- | 13.2<br>19<br>24 | 6.6<br>9<br>12 | 5.3<br>8<br>10 | -- |
| 15<br>20<br>25 | -- | 36<br>47<br>59 | 18<br>23<br>29 | 14<br>19<br>24 | -- |
| 30<br>40<br>50 | -- | 69<br>90<br>113 | 35<br>45<br>56 | 28<br>36<br>45 | -- |
| 60<br>75<br>100 | -- | 133<br>166<br>218 | 67<br>83<br>109 | 53<br>66<br>87 | 14<br>18<br>23 |
| 125<br>150<br>200 | -- | 270<br>312<br>416 | 135<br>156<br>208 | 108<br>125<br>167 | 28<br>32<br>43 |

Table 4.12

| Full-load Current Values for Induction Type, Squirrel-Cage & Wound-Rotor 3-Phase AC Motors (In Amperes) | | | | | | | |
|---|---|---|---|---|---|---|---|
| Hp | 115 Volts | 200 Volts | 208 Volts | 230 Volts | 460 Volts | 575 Volts | 2300 Volts |
| 1/2 | 4.4 | 2.5 | 2.4 | 2.2 | 1.1 | 0.9 | |
| 3/4 | 6.4 | 3.7 | 3.5 | 3.2 | 1.6 | 1.3 | -- |
| 1 | 8.4 | 4.8 | 4.6 | 4.2 | 2.1 | 1.7 | |
| 1 1/2 | 12.0 | 6.9 | 6.6 | 6.0 | 3.0 | 2.4 | |
| 2 | 13.6 | 7.8 | 7.5 | 6.8 | 3.4 | 2.7 | -- |
| 3 | | 11.0 | 10.6 | 9.6 | 4.8 | 3.9 | |
| 5 | | 17.5 | 16.7 | 15.2 | 7.6 | 6.1 | |
| 7 1/2 | -- | 25.3 | 24.2 | 22 | 11 | 9 | -- |
| 10 | | 32.2 | 30.8 | 28 | 14 | 11 | |
| 15 | | 48.3 | 46.2 | 42 | 21 | 17 | |
| 20 | -- | 62.1 | 59.4 | 54 | 27 | 22 | -- |
| 25 | | 78.2 | 74.8 | 68 | 34 | 27 | |
| 30 | | 92 | 88 | 80 | 40 | 32 | |
| 40 | -- | 120 | 114 | 104 | 52 | 41 | -- |
| 50 | | 150 | 143 | 130 | 65 | 52 | |
| 60 | | 177 | 169 | 154 | 77 | 62 | 16 |
| 75 | -- | 221 | 211 | 192 | 96 | 77 | 20 |
| 100 | | 285 | 273 | 248 | 124 | 99 | 26 |
| 125 | | 359 | 343 | 312 | 156 | 125 | 31 |
| 150 | -- | 414 | 396 | 360 | 180 | 144 | 37 |
| 200 | | 552 | 528 | 480 | 240 | 192 | 49 |
| 250 | | | | | 302 | 242 | 60 |
| 300 | -- | -- | -- | -- | 361 | 289 | 72 |
| 350 | | | | | 414 | 336 | 83 |
| 400 | | | | | 477 | 382 | 95 |
| 450 | -- | -- | -- | -- | 515 | 412 | 103 |
| 500 | | | | | 590 | 472 | 118 |

Stated values typical for motors running at speeds usual for belted motors and for those with normal torque characteristics. More current may be required for motors designed for low speeds of 1200 RPM or less and for those with high torque. For multi-speed motors, full-load current will vary with speed; in these cases, current rating on the nameplate must be used. Voltages listed are rated voltages. Currents given shall be permitted for system voltage ranges of 110-120, 220-240, 440-480, and 550-600 volts.

Table 4.13

| Full-Load Current Values for 3-Phase AC Motors | | | | |
|---|---|---|---|---|
| Synchronous Type Unity Power Factor* In Amperes | | | | |
| **Hp** | **230 Volts** | **460 Volts** | **575 Volts** | **2300 Volts** |
| 25 | 53 | 26 | 21 | -- |
| 30 | 63 | 32 | 26 | -- |
| 40 | 83 | 41 | 33 | -- |
| 50 | 104 | 52 | 42 | -- |
| 60 | 123 | 61 | 49 | 12 |
| 75 | 155 | 78 | 62 | 15 |
| 100 | 202 | 101 | 81 | 20 |
| 125 | 253 | 126 | 101 | 25 |
| 150 | 302 | 151 | 121 | 30 |
| 200 | 400 | 201 | 161 | 40 |

*For 90% power factor, multiply above values by 1.1; for 80% power factor, multiply by 1.25. Stated values typical for motors running at speeds usual for belted motors and for those with normal torque characteristics. More current may be required for motors designed for low speeds of 1200 RPM or less and for those with high torque. For multi-speed motors, full-load current will vary with speed; in these cases, current rating on the nameplate must be used. Voltages listed are rated voltages. Currents given shall be permitted for system voltage ranges of 110-120, 220-240, 440-480, and 550-600 volts.

Table 4.14

| Maximum Rating/Setting of Short-circuit & Ground-fault Protective Devices for Motor Branch Circuits | | | | |
|---|---|---|---|---|
| Percentage of Full-load Current | | | | |
| **Motor Type** | **Nontime Delay Fuse**** | **Dual Element (Time Delay Fuse)**** | **Instantaneous Trip Breaker** | **Inverse Time Breaker*** |
| Single-phase | 300 | 175 | 800 | 250 |
| **Polyphase AC Motors Except Wound Rotor** | | | | |
| Squirrel Cage: Design E | 300 | 175 | 1100 | 250 |
| Other than Design E | 300 | 175 | 800 | 250 |
| Synchronous | 300 | 175 | 800 | 250 |
| **Other Motor Types** | | | | |
| Wound Rotor | 150 | 150 | 800 | 150 |
| Direct current (constant voltage) | 150 | 150 | 250 | 150 |

*Values in last column also cover ratings of nonadjustable inverse time types of circuit breakers that may be modified as in NEC 430.52.

**Values in the Nontime Delay Fuse column also apply to Time Delay Class CC fuses.

Low-torque, low-speed (usually 450 rpm or lower) synchronous motors like those used to drive reciprocating compressors, pumps, etc., that start unloaded, do not require a fuse rating or circuit breaker setting greater than 200% of full-load current.

The National Electrical Code requires the use of code letters on motor nameplates to show motor input with locked rotor. Table following shows the letter designations for various motor inputs.

Table 4.15

| Code Letters Indicating Locked-Rotor Motor Input | | |
|---|---|---|
| **Code Letter** | **Kilovolts** | **Amperes** |
| A | 0 | 3.14 |
| B | 3.15 | 3.54 |
| C | 3.55 | 3.99 |
| D | 4.0 | 4.49 |
| F | 5.0 | 5.59 |
| G | 5.6 | 6.29 |
| H | 6.3 | 7.09 |
| J | 7.1 | 7.99 |
| K | 8.0 | 8.99 |
| L | 9.0 | 9.99 |
| M | 10.0 | 11.19 |
| N | 11.2 | 12.49 |
| P | 12.5 | 13.99 |
| R | 14.0 | 15.99 |
| S | 16.0 | 17.99 |
| T | 18.0 | 19.99 |
| U | 20.0 | 22.39 |
| V | 22.4 | and up |

Table 4.16

| Full-load Current in Amperes for DC Motors* Armature Voltage Rating** | | | | | | |
|---|---|---|---|---|---|---|
| Hp | 90 V | 120 V | 180 V | 240 V | 500 V | 550 V |
| 1/4 | 4.0 | 3.1 | 2.0 | 1.6 | -- | -- |
| 1/3 | 5.2 | 4.1 | 2.6 | 2.0 | | |
| 1/2 | 6.8 | 5.4 | 3.4 | 2.7 | | |
| 3/4 | 9.6 | 7.6 | 4.8 | 3.8 | -- | -- |
| 1 | 12.2 | 9.5 | 6.1 | 4.7 | | |
| 1 1/2 | -- | 13.2 | 8.3 | 6.6 | | |
| 2 | -- | 17 | 10.8 | 8.5 | -- | -- |
| 3 | | 25 | 16 | 12.2 | | |
| 5 | | 40 | 27 | 20 | | |
| 7 1/2 | -- | 58 | -- | 29 | 13.6 | 12.2 |
| 10 | | 76 | | 38 | 18 | 16 |
| 15 | | -- | | 55 | 27 | 24 |
| 20 | -- | -- | -- | 72 | 34 | 31 |
| 25 | | | | 89 | 43 | 38 |
| 30 | | | | 106 | 51 | 46 |
| 40 | -- | -- | -- | 140 | 67 | 61 |
| 50 | | | | 173 | 83 | 75 |
| 60 | | | | 206 | 99 | 90 |
| 75 | -- | -- | -- | 255 | 123 | 111 |
| 100 | | | | 341 | 164 | 148 |
| 125 | | | | 425 | 205 | 185 |
| 150 | -- | -- | -- | 506 | 246 | 222 |
| 200 | | | | 675 | 330 | 294 |

*These values are for motors running at base speed.
** Armature voltage rating.

## Power Factor

Due to induction that is present in certain types of AC circuits, such as motor and starter coils, current will lag the voltage and thus be out of phase. That is, the current will not reach its maximum value at the same time that the voltage reaches its maximum value. This out-of-phase difference is measured in degrees and is called **phase angle.**

As a result of phase angle, the power in a three-phase AC circuit must be calculated using the following formula:

$$P = I \times E \times \textit{Cosine of Phase Angle} \times 1.73$$

This power, measured in watts, is called true power.

**For example:**

A 15 hp motor draws 42 A at 230 V three-phase with a phase angle of 30 degrees. Its true power would be:

$$P = I \times E \times COS\ 30^\circ \times 1.73$$

$$= 42 \times 230 \times .866 \times 1.73$$

$$= 14{,}472\ \textit{watts}\ (14.47\ kW)$$

However, the actual volt-amperes supplied through the circuit to this motor are higher:

$$P = I \times E \times 1.73 \text{ (for three circuits)}$$

$$= 42 \times 230 \times 1.73$$

$$= 16{,}712\ \textit{volt-amperes}\ (16.71\ kVA)$$

This is known as apparent power.

The ratio of the true power (14.47 kW) to the apparent power (16.71 kVA) is known as power factor and is given as a percentage. The ratio is expressed with the formula:

$$Power\ Factor = \frac{True\,Power}{Apparent\,Power}$$

In our example: $PF = \frac{14.47}{16.71} = .866\ (86.6\%)$

The difference between true power and apparent power represents lost energy. This loss can be corrected by the use of capacitors installed in the supply lines.

**Electric Motor Efficiency**

The efficiency of any motor is based on the power output compared to power input.

$$\%\ \text{Eff.} = \frac{Output}{Input} x\, 100$$

**Example 1:**

What is the efficiency of a motor that uses 4.5 kilowatts and produces 5 horsepower?

**Solution:** 1 HP = 746 watts

5 HP = 5 x 746 watts = 3,730 watts

4.5 kilowatts = 4500 watts

$$\%\ \text{Eff.} = \frac{3730\ watts}{4500\ watts}$$

$$= .829 = 83\%$$

**Example 2:**

A motor draws 45 kilowatts and has an output of 55 horsepower. What is this motor's efficiency?

**Solution:** 1 HP = 746 watts

55 HP = 55 X 746 watts = 41,030 watts

45 kW = 45,000 watts

$$\% \text{ Eff.} = \frac{41{,}030 \; watts}{45{,}000 \; watts}$$

$$= .912 = 91.2\%$$

Table 4.17

| **Motor Running Overload Protection** | | | |
|---|---|---|---|
| **Motor Type** | **Circuit Supply** | **Number of Overload Units** | **Location of Overload Units** |
| Single-Phase AC or DC | 2-Wire, Single-Phase Ungrounded | One | Either Conductor |
| Single-Phase AC or DC | 2-Wire Single-Phase, 1 Conductor Grounded | One | In the Ungrounded Conductor |
| Single-Phase AC or DC | 3-Wire Single-Phase Grounded Neutral | One | In Either Ungrounded Conductor |
| Three-Phase AC | Any Type Three-Phase Supply | Three | One in Each Phase, Unless Protected by Another Approved Means |

Figure 4.6

**When locating the motor control center, place it in sight of the motors.**

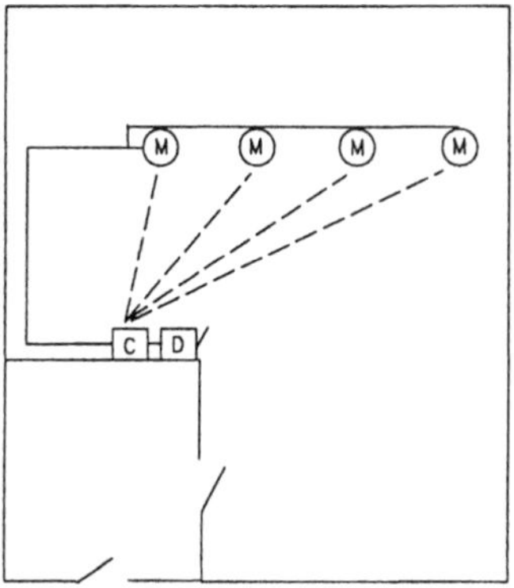


Figure 4.7

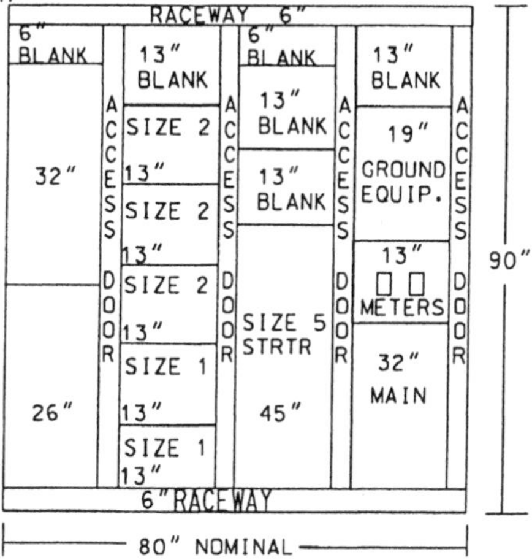


**Typical motor control center.**

Figure 4.8

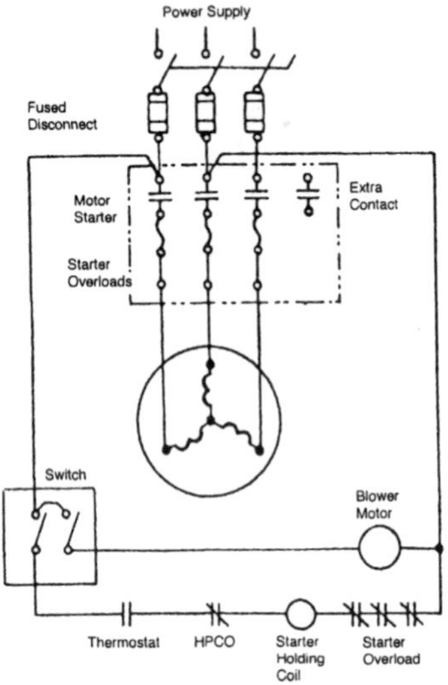


**Typical motor control. Typical system wiring diagram shows a 4-pole motor starter.**

Figure 4.9

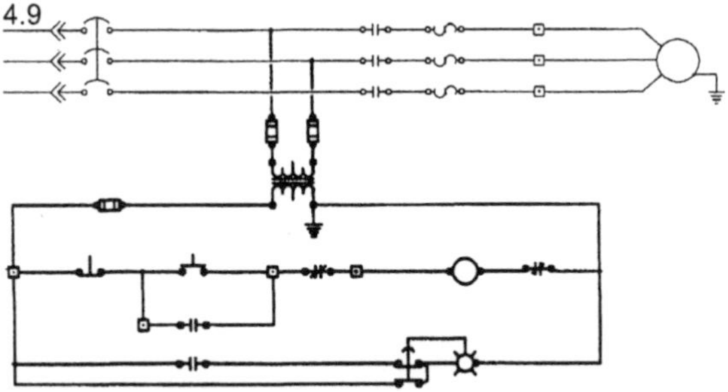

**Shown at top in finer lines: 3-phase motor control wiring diagram; in darker lines: control schematic with remote stop-start device.**

Figure 4.10

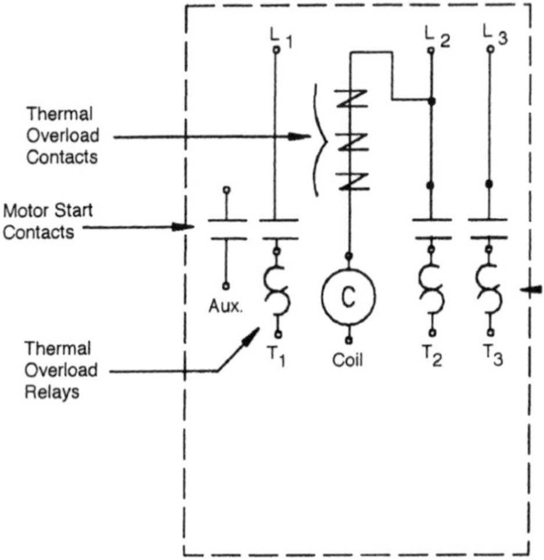


**Magnetic starter showing its parts.**

Figure 4.11

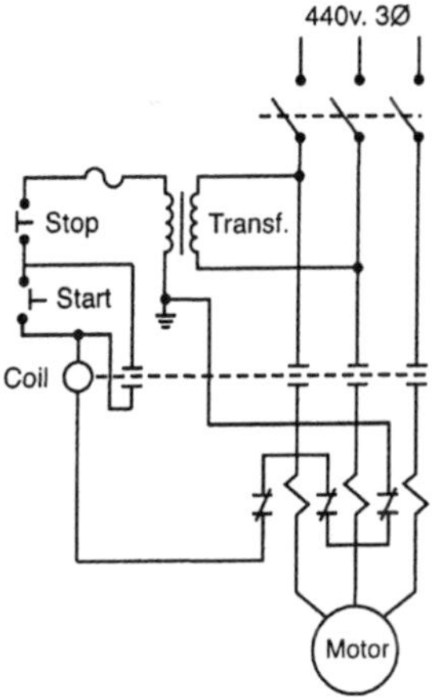


**Schematic of typical motor starter.**

Table 4.18

| Checklist for Once-a-Year Motor Maintenance | |
|---|---|
| **Steps** | **Maintenance Work to Be Done** |
| 1 | Turn off and lock out power to motor and its control circuit. |
| 2 | Clean outside of motor and ventilation ducts. |
| 3 | Disconnect motor from load and disassemble. |
| 4 | Clean inside of motor. |
| 5 | Check centrifugal switch assembly. |
| 6 | Check rotor, armature, and field windings. |
| 7 | Check all peripheral equipment. |
| 8 | Check bearings for wear and defects. |
| 9 | Check brushes and commutator. |
| 10 | Check slip rings. |
| 11 | Reassemble motor and couple to load. |
| 12 | Flush away old bearing lubricant and replace. |

Table 4.18 (cont.)

| Checklist for Once-a-Year Motor Maintenance | |
|---|---|
| **Steps** | **Maintenance Work to Be Done** |
| 13 | Check motor's wiring raceway. |
| 14 | Check drive mechanism. |
| 15 | Check motor terminals. |
| 16 | Check capacitors. |
| 17 | Check mounting bolts for tightness. |
| 18 | Check and record line-to-line resistance. |
| 19 | Check and record Megohmmeter resistance from $T_1$ to ground. |
| 20 | Check and record insulation polarization index. |
| 21 | Check motor controls. |
| 22 | Reconnect motor and control circuit power supplies. |
| 23 | Check line-to-line voltage for balance and level. |
| 24 | Check line current draw against nameplate rating. |
| 25 | Check and record inboard and outboard bearing temperatures. |

Table 4.19

| Electric Motor Maintenance Checklist-Semiannual | |
|---|---|
| **Sequence** | **Maintenance Operation To Be Performed** |
| Step 1 | Turn off and lock out all electrical power to the motor and its control circuit. |
| Step 2 | Thoroughly clean outside of motor and its ventilation ducts. |
| Step 3 | Check the motor's conductor raceway. |
| Step 4 | Examine and lubricate bearings as needed. |
| Step 5 | Check drive pulley and shaft. |
| Step 6 | Check brushes and commutator for wear and other problems. |
| Step 7 | Check slip rings. |
| Step 8 | Check motor terminals. |
| Step 9 | Examine capacitors. |
| Step 10 | Check mounting bolts for tightness. |
| Step 11 | Check and record line-to-line resistance. |
| Step 12 | Note and record meghommeter resistance from $L_1$ to ground. |
| Step 13 | Check the motor controls. |
| Step 14 | Restore power to motor and circuit power supplies. |
| Step 15 | Check line-to-line voltage for level and balance. |
| Step 16 | Note nameplate rating and check for proper line current draw. |
| Step 17 | Check and make a record of both inboard and outboard bearing temperatures. |

Table 4.20

| Electric Motor Troubleshooting | | |
|---|---|---|
| **Problem** | **Cause(s)** | **Remedies** |
| Motor gets too hot | Rotor rubbing on stator from bent shaft or worn bearings | Replace bent shaft or worn bearings. |
| | Overloading or binding load | Check and correct overload. Check the current input. |
| | Poor ventilation | Clean off the motor. Clean vent holes or venting system. |
| No start | Power failure: voltage too low | Check circuit breakers or fuses; replace as needed. |
| | Improper connections | Study wiring diagram and make any needed corrections. |
| | Overload | Reduce load or install a larger motor. |
| | Brushes worn or set improperly | Replace or reset brushes. |
| | Open circuit | Recheck all connections. Clean contacts on starting switch. Look for short-circuits or ground faults. Reset tripped thermal overload. |
| | Excess end play | Add washers to shaft. |

Table 4.20 (cont.)

| Electric Motor Troubleshooting | | |
|---|---|---|
| **Problem** | **Cause(s)** | **Remedies** |
| Rotor or stator burns out | Moisture, corrosive chemicals or dust present | Clean and dry motor; shield it, if necessary. |
| Brushes sparking | Short-circuit or open circuit | Clean and repair or replace armature. |
| | Brushes worn or sticking | Replace brushes. |
| | Dust in motor | Clean motor. |
| | Overloaded motor | Reduce the load or use a larger motor. |
| | Brushes improperly fitted | Refit them to match the commutator. |
| | Commutator is loose | Replace it. |
| Motor noisy | Load unbalanced | Balance load and pulley. |
| | Parts may be loose | Tighten motor parts and motor mounts. |
| | Bad alignment | Realign motor with load. |
| | Shaft is bent | Straighten or replace & align. Make sure load is balanced. |
| | Bearings worn | Lubricate or replace them. |
| | Dusty | Clean motor. |
| Good maintenance is important; without it a motor's life is shortened greatly. | | |

Table 4.21

| Symbols Used in Wiring Diagrams for Electric Motor Starters | | |
|---|---|---|
| **Device** | **Type** | **Symbol** |
| **Coils** | Relay and Switch Coils |  Single Winding 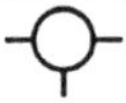 Tapped  Permanent Magnet  Economized |
| **Contacts** | Normally Closed (N.C.) |  Main 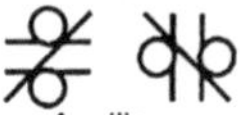 Auxiliary |
| | Normally Open (N.O.) | 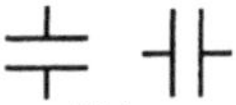 Main 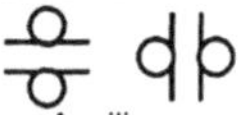 Auxiliary |
| | Time Closing |  N.O.T.C. 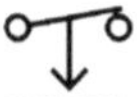 N.C.T.C. |
| | Time Opening |  N.C.T.O. 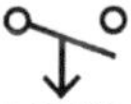 N.O.T.O. |
| **Contactors** | AC Solenoid Type | 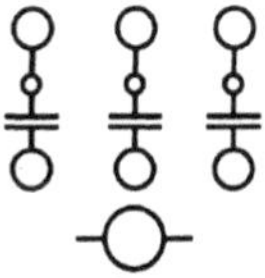 3 Pole |
| | Manually Operated | 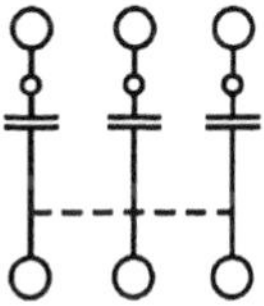 |

Table 4.21 (cont.)

| Symbols Used in Wiring Diagrams for Electric Motor Starters cont. | | |
|---|---|---|
| **Device** | **Type** | **Symbol** |
| **Fuse** | General |  |
| **Indicating Lights** | General |   NE - Neon<br>FL - Flourescent<br>P - Purple<br>OP - Opalescent<br>O - Orange<br>A - Amber<br>B - Blue<br>C - Clear<br>G - Green<br>R - Red<br>W - White<br>Y - Yellow |
| **Motors** | 3-Phase Squirrel Cage Induction | 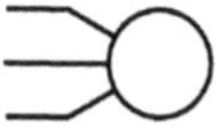 |
| | Single Phase | 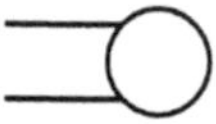 |
| **Rectifier** | Full Wave with Color Code | 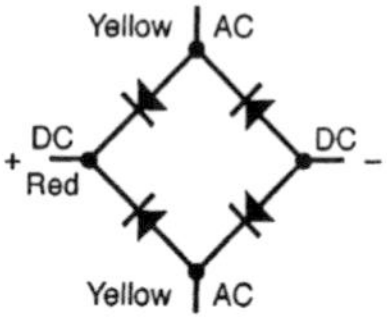  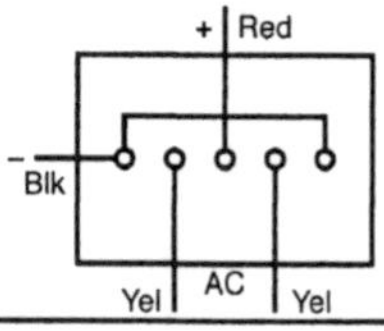  |

Table 4.21 (cont.)

| Symbols Used in Wiring Diagrams for Electric Motor Starters cont. | | |
|---|---|---|
| **Device** | **Type** | **Symbol** |
| **Relays** | Control | 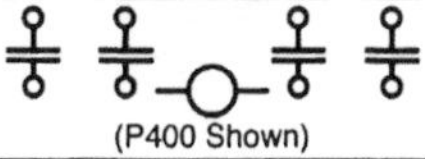(P400 Shown) |
| | Thermal Overload | 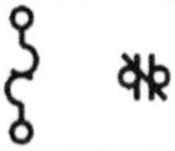 |
| | Timing (Pneumatic) (ON-DELAY) | 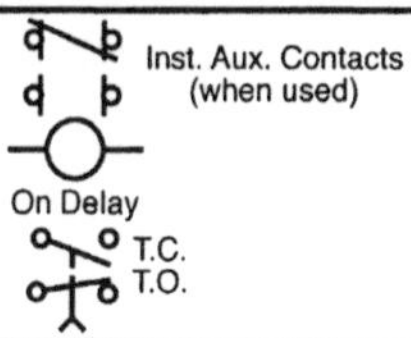Inst. Aux. Contacts (when used)<br>On Delay<br>T.C.<br>T.O. |
| **Switches** | Anti-Plugging | 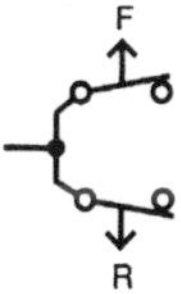F<br>R |
| | Float Switch | 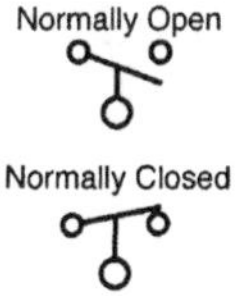Normally Open<br>Normally Closed |
| | Limit Switches | 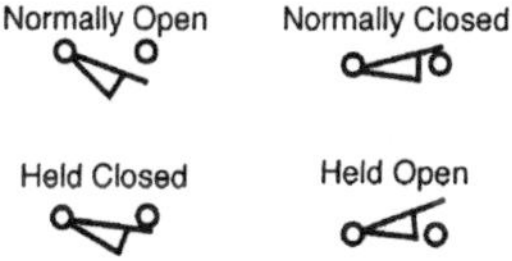Normally Open<br>Normally Closed<br>Held Closed<br>Held Open |

Table 4.21 (cont.)

| Symbols Used in Wiring Diagrams for Electric Motor Starters cont. | | |
|---|---|---|
| **Device** | **Type** | **Symbol** |
| **Switches** | Plugging | 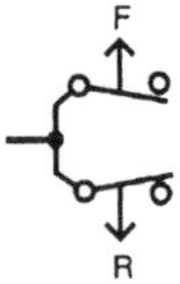 |
| | Pressure and Temperature | 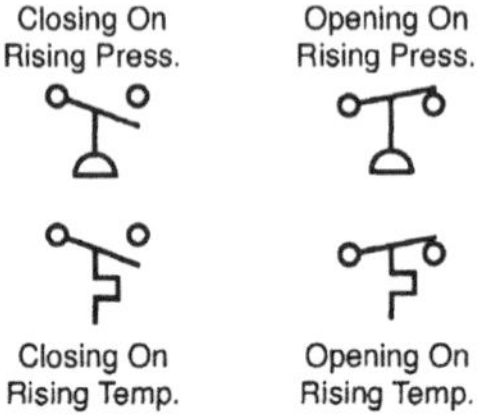Closing On Rising Press. Opening On Rising Press. Closing On Rising Temp. Opening On Rising Temp. |
| | Standard Push Button | NC NO |
| | Heavy-Duty, Oil-tight Push Button | Mushroom Head |
| | Push Button and Jog Attachment | 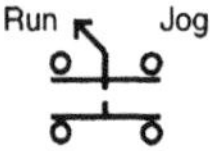Run Jog |
| | Standard Duty Selector Switch | 2 Position 3 Position |

Table 4.21 (cont.)

## Symbols Used in Wiring Diagrams for Electric Motor Starters cont.

<table>
<tr><th>Device</th><th>Type</th><th>Symbol</th></tr>
<tr><td rowspan="2">Switches</td><td>Heavy Duty Selector 2-Position</td><td>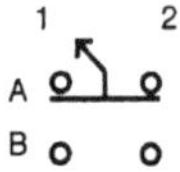

<table>
<tr><td rowspan="2">Letter Sym.</td><td colspan="2">Position</td></tr>
<tr><td>1</td><td>2</td></tr>
<tr><td>A</td><td>X</td><td></td></tr>
<tr><td>B</td><td></td><td>X</td></tr>
</table></td></tr>
<tr><td>Heavy Duty Selector 3-Position</td><td>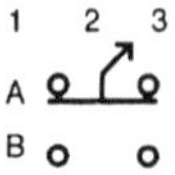

<table>
<tr><td rowspan="2">Letter Sym.</td><td colspan="3">Position</td></tr>
<tr><td>1</td><td>2</td><td>3</td></tr>
<tr><td>A</td><td></td><td></td><td>X</td></tr>
<tr><td>B</td><td>X</td><td></td><td></td></tr>
</table></td></tr>
<tr><td rowspan="2">Trans-formers</td><td>Potential</td><td>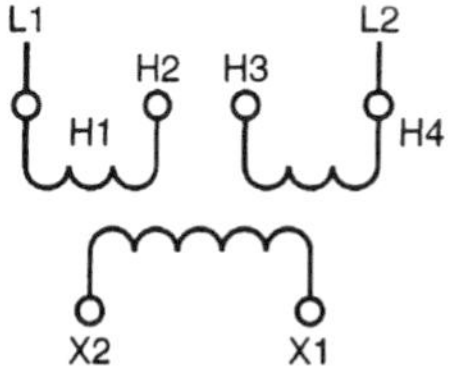
</td></tr>
<tr><td>Current</td><td>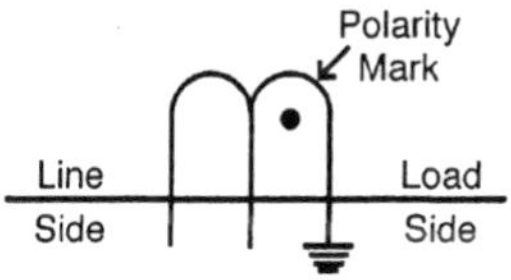
</td></tr>
</table>

# UNIT 5

# TRANSFORMERS

Transformers, found only in AC circuits, link two different circuits without using any physical connection between them. Transformers are necessary in the distribution of alternating current electric power since they can convert that power from a given current and voltage to a lower or higher current and voltage.

Linkage is possible through inductance. That is, current and voltage in a coil connected to the applied voltage in the source circuit set up magnetic lines of force (flux). When the flux lines from one coil cut the windings of another nearby coil, a voltage will be induced in that coil. The coil in which the flux is induced by a power source is called the primary. The one in which the voltage is induced by the magnetic field of the primary is called the secondary. The primary coil couples its energy with the secondary winding by way of a changing magnetic field.

The difference in the number of windings between the primary and secondary determine the difference in current and voltage between them.

Figure 5.1

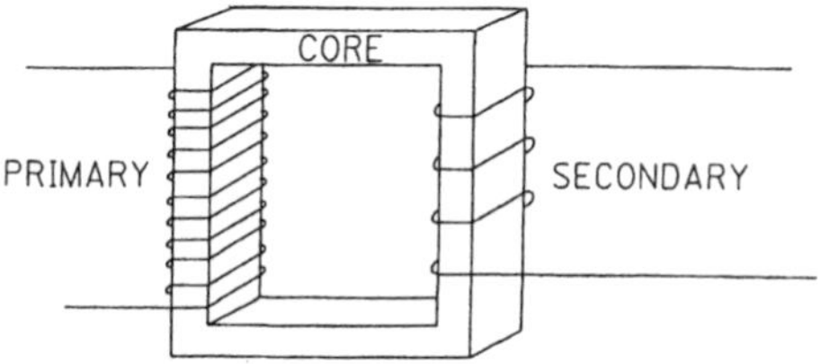


**Basic makeup of a transformer. Essentially, it is two coils of wire wrapped around an iron core, although not all transformers are made that way. There is no physical connection between the two coils. Note that there are more windings on the primary coil than on the secondary coil.**

Figure 5.2

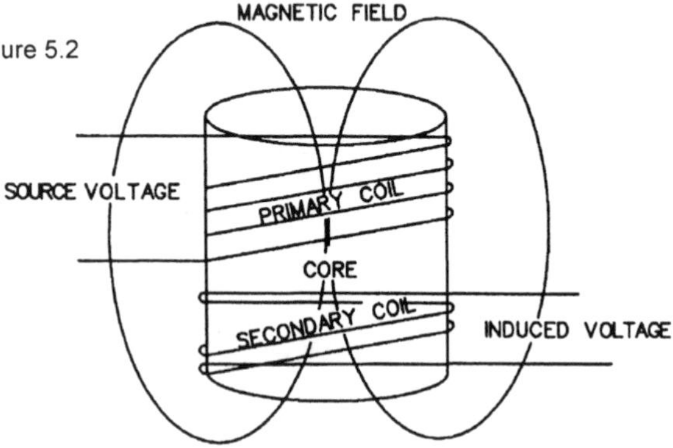


**Basic action of a transformer. When voltage is present in the primary circuit from its power source and a load is connected to the secondary, voltage is induced in the secondary through action of the magnetic field set up by the coils.**

This is how a transformer works:

1. When there is no current in the secondary circuit (as when there is no load connected to it) there is little current in the primary. This is due to the back electromotive force created by the primary coil.

2. When a load is connected to the secondary circuit, there is current in the secondary and current in the primary increases in direct proportion to the secondary current. This occurs because the magnetic field in the secondary coil opposes the magnetic field in the primary, thus weakening the back electromotive force.

Since current in the primary increases in proportion to the current in the secondary, it is easy to see what would happen in case of a short in the secondary circuit. Not only would the transformer burn out but the source supplying power to the primary would be damaged.

Figure 5.3

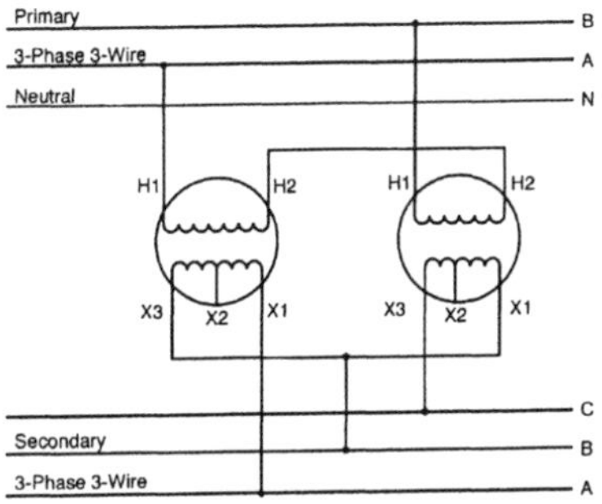


**Open Y-Delta - In this connection, the system is unbalanced and one phase is disabled.**

Figure 5.4

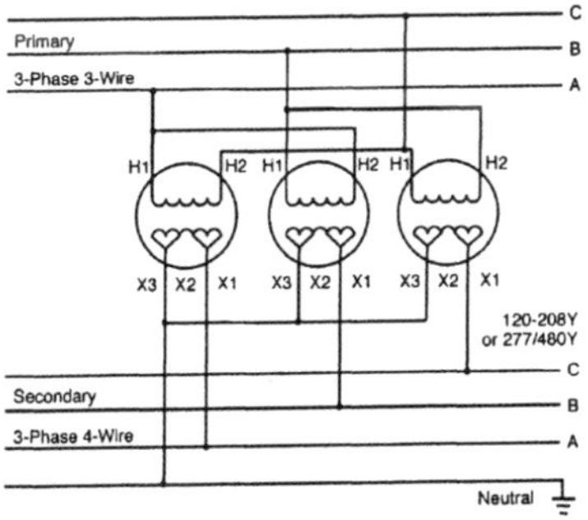


**Delta-Y for Lighting and Power - In this connection, the single-phased loads can be balanced and the banks can be paralleled.**

Figure 5.5

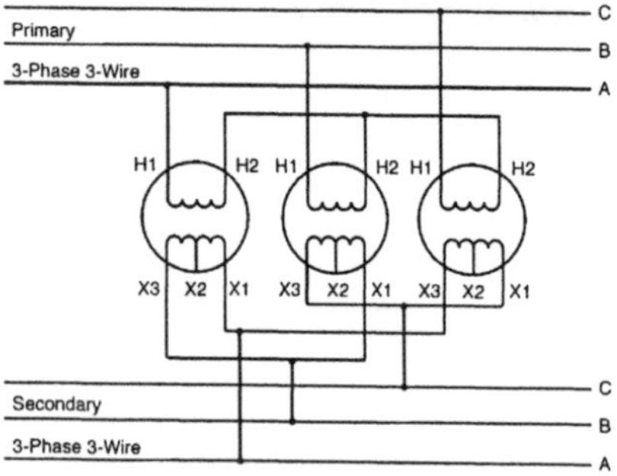


**Y-Delta Power Connection - This connection is used to increase voltage from 2400 to 4160 volts. The primary neutral is not grounded or tied to this system.**

Figure 5.6

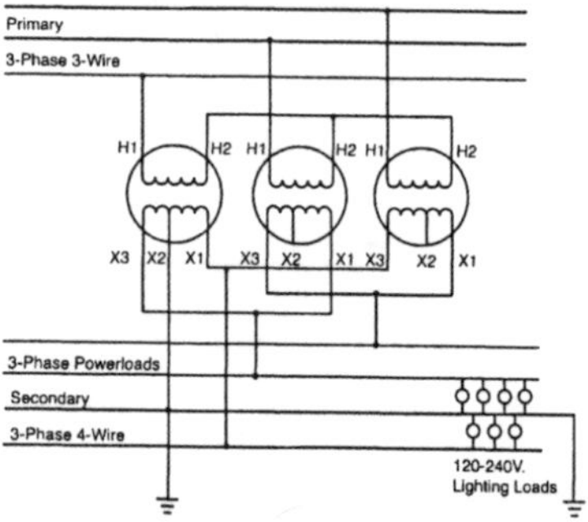


**Y-Delta for Lighting and Power - Diagram shows a Y-Delta bank supplying both lighting and power. The primary neutral is not grounded or tied to the system neutral. This connection requires special watt-hour metering.**

Figure 5.7

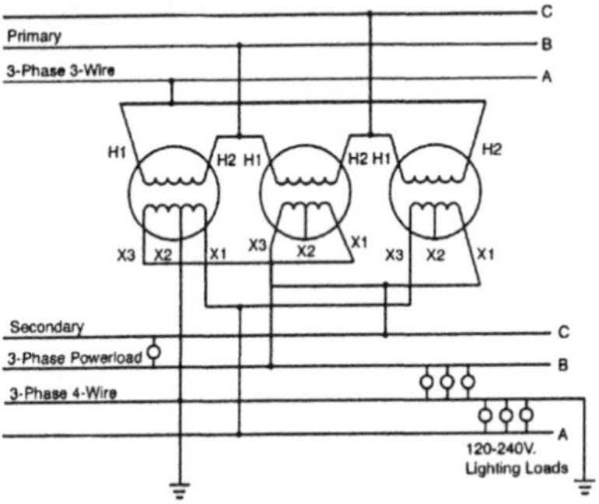


**Delta-Delta for Lighting and Power - This connection is often used to supply small single-phase lighting and 3-phase power loads at the same time. This connection is not available from all utility companies.**

Figure 5.8

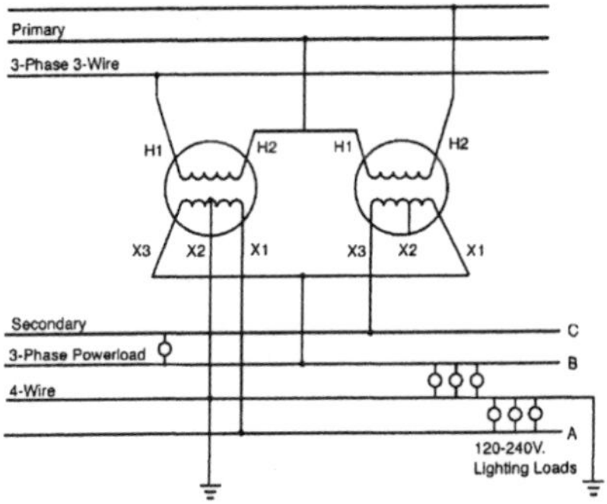


**Open Delta for Lighting and Power - When the secondary load is a combination of single-phase and 3-phase power, this connection is frequently used. It is particularly beneficial when the single-phase lighting load is large and the 3-phase power load is small.**

Figure 5.9

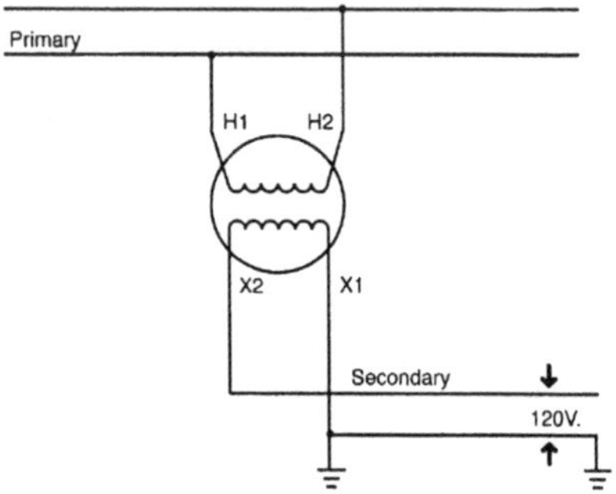


**Single-phase 120 volt Lighting - This connection is used where the load is small and the length of the secondary circuit is short.**

Figure 5.10

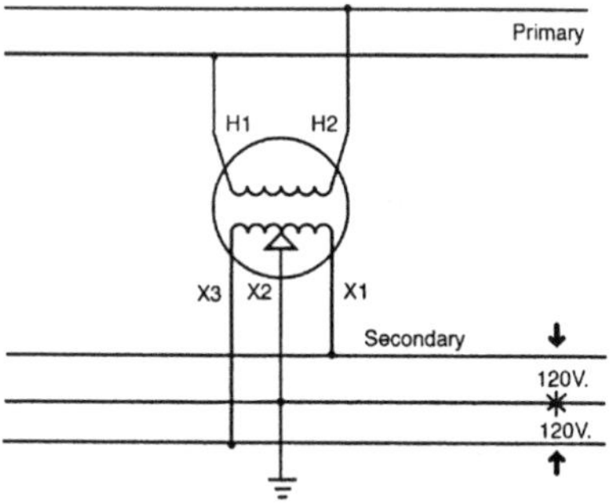


**Single-phase 120/240, 3-wire Lighting/Power Load - This connection serves both 120 and 240 volt loads simultaneously.**

Figure 5.11

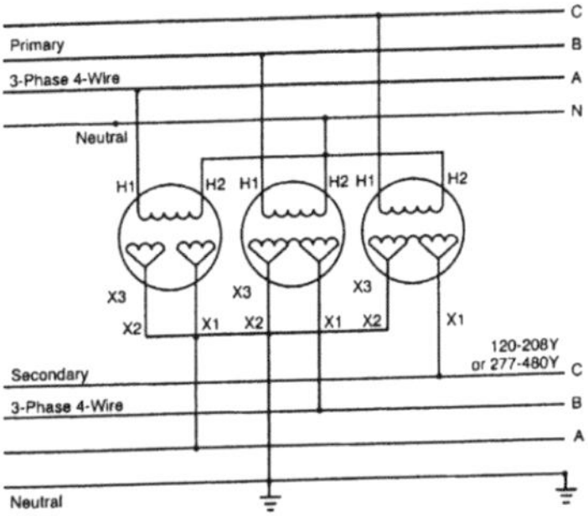


**Y-Y for Lighting and Power - This diagram shows a system where the primary voltage was increased from 2400 volts to 4160 volts. This arrangement increases the potential capacity of the system.**

Figure 5.12

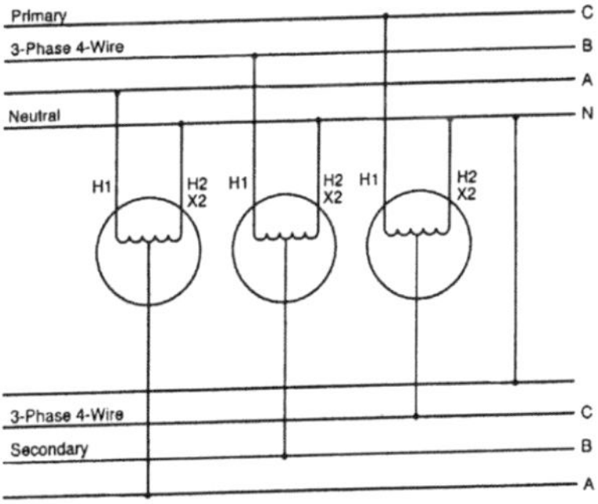


**Y-Y Autotransformer for Supplying Power from a 3-Phase, 4-Wire System - This connection is the most economical way of stepping down voltage when the ratio of primary to secondary voltage is small. Branch circuits must not be supplied by autotransformer.**

Figure 5.13

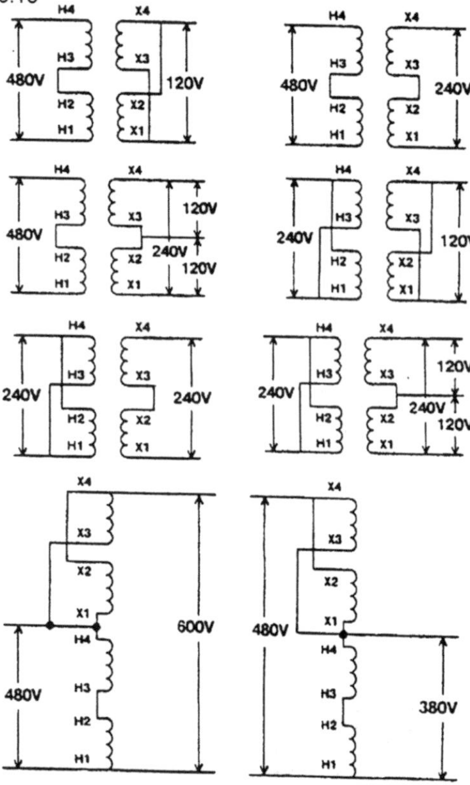


Output kVA = 5 x Nameplate Rating Autotransformer Connection

Output kVA = 4 x Nameplate Rating Autotransformer Connection

**Diagrams for different arrangements of single-phase transformers.**

Figure 5.14a

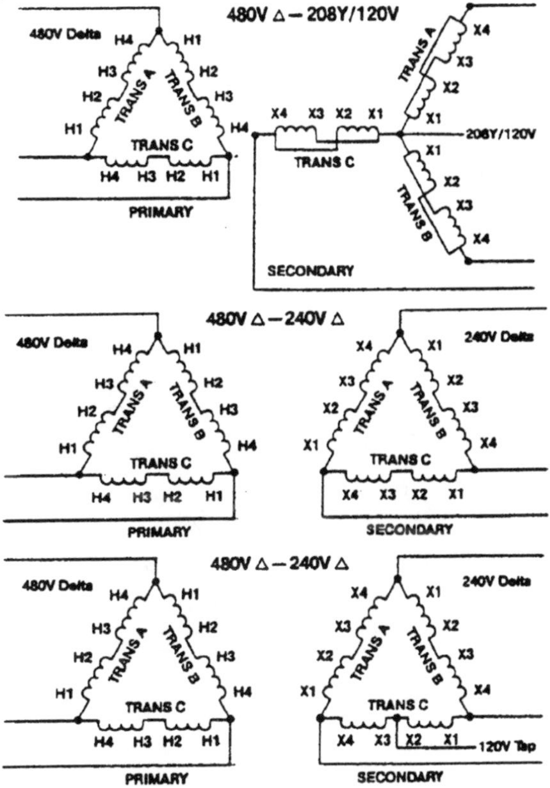


Three wiring diagrams for banking single-phase transformers. These three arrangements are for 3-phase operation. Total kVA output is the sum of the three transformers.

Figure 5.14b

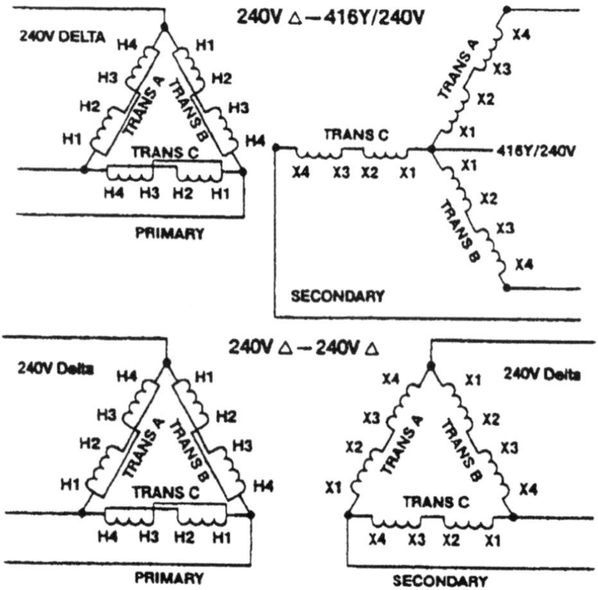


**More diagrams showing connections for single-phase transformers banked for three-phase operation.**

Figure 5.14c

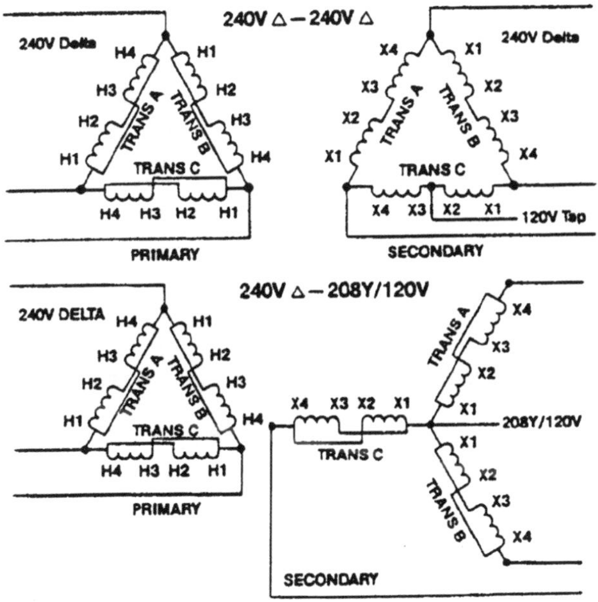


Additional diagrams of single-phase transformers banked for three-phase operation.

Figure 5.15

| Bestever Transformer Corporation<br>12345 Main St.<br>Anytown, USA | |
|---|---|
| kVa: 50 | Single Phase |
| HV Winding: 2400 V | Low V. Winding: 240 V |
| Freq.: 60 HZ | % Z = 3.5 |
| Polarity: Additive | Temp. Rise: 55 Degrees |
| Serial # 1234567 | XXX Gallons Oil |
| Model # 01010101 | Type: ZZZ |

**Typical nameplate required for all transformers. It is located anywhere on the transformer case. Polarity of any transformer is merely an indication of the direction of current flow at any given moment. This polarity is additive when you connect the adjacent high and low terminals. This has the effect of increasing voltage by the sum of the high and low voltage windings.**

Figure 5.16

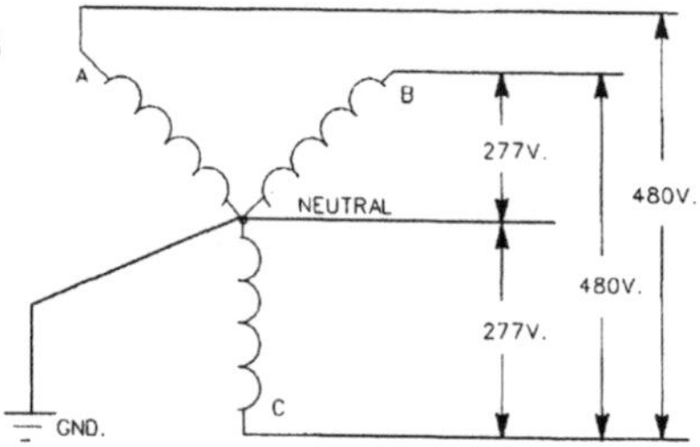


**480/277 V four-wire, 3-phase Distribution System - This wye arrangement is typical of commercial wiring, such as large factories, commercial buildings, schools, shopping malls, and high-rise buildings where 277 V fluorescent lighting and 480 V motors are often the preferred voltage ratings.**

Figure 5.17

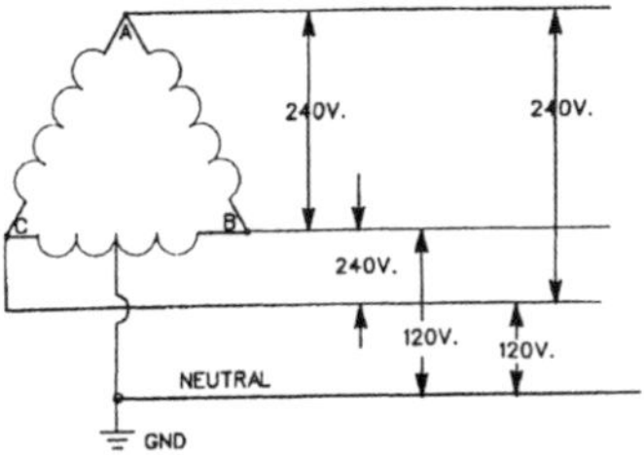


**A delta transformer arrangement commonly used in commercial and industrial facilities. It supplies 3-phase power, at 240 and 120 volts using four wires.**

## GROUNDING OF SECONDARY IN A TRANSFORMER

Grounding of the secondary coil of a transformer (see illustration following) is the same as the Code requirement for the grounding of an AC system. Size of the electrode conductor should never be less than that shown in the table below. There are some more rules to be found in the following Code sections: 250.66(A), 250.66(B), and 250.66(C).

Figure 5.18

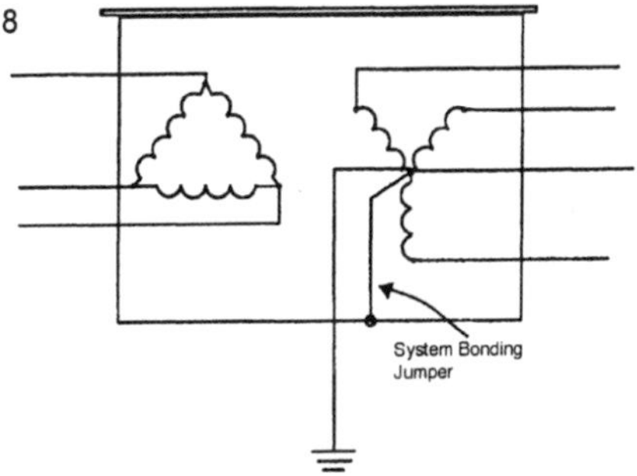

Transformer secondary grounding electrode conductor and bonding jumper sizes can be determined from the table following:

Table 5.1

| Determining Sizes of Grounding Electrode Conductor and System Bonding Jumper for Transformer Secondary** | | | |
|---|---|---|---|
| Size: Largest Service Entrance Conductor or Equivalent Area for Parallel Conductors | | Size: Grounding Electrode Conductor | |
| Copper | Aluminum or Copper-Clad Aluminum | Copper | Aluminum or Copper-Clad Aluminum |
| 2 or smaller | 1/0 or smaller | 8 | 6 |
| 1 or 1/0 | 2/0 or 3/0 | 6 | 4 |
| 2/0 or 3/0 | 4/0 or 350 kcmil | 4 | 2 |
| Over 3/0 thru 350 kcmil | Over 250 kcmil thru 500 kcmil | 2 | 1/0 |
| Over 350 kcmil thru 600 kcmil | Over 500 kcmil thru 900 kcmil | 1/0 | 3/0 |
| Over 600 kcmil thru 1100 kcmil | Over 900 kcmil thru 1750 kcmil | 2/0 | 4/0 |
| Over 1100 kcmil | Over 1750 kcmil | 3/0 | 250 kcmil |
| **See NEC Table 250.66 | | | |

Figure 5.19

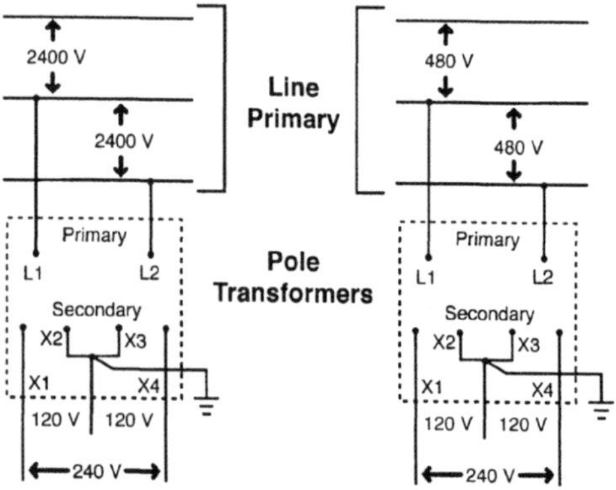


**Transformer wiring diagram for two pole transformers showing connections to primary.**

Table 5.2

| Three-Phase Voltage (Line to Line) Full-Load Current | | | | | | |
|---|---|---|---|---|---|---|
| kVA Rating | **208** | **240** | **480** | **2400** | **4160** | **13200** |
| 3 | 8.3 | 7.2 | 3.6 | .72 | .415 | .13 |
| 6 | 16.6 | 14.4 | 7.2 | 1.44 | .83 | .26 |
| 9 | 25 | 21.6 | 10.8 | 2.16 | 1.25 | .40 |
| 15 | 41.6 | 36.0 | 18.0 | 3.6 | 2.1 | .65 |
| 30 | 83 | 72 | 36 | 7.2 | 4.15 | 1.3 |
| 45 | 125 | 108 | 54 | 10.8 | 6.25 | 1.97 |
| 75 | 208 | 180 | 90 | 18 | 10.4 | 3.3 |
| 100 | 278 | 241 | 120 | 24 | 13.9 | 4.4 |
| 150 | 416 | 360 | 180 | 36 | 20.8 | 6.55 |
| 225 | 625 | 542 | 271 | 54 | 31.2 | 9.85 |
| 300 | 830 | 720 | 360 | 72 | 41.5 | 13.1 |
| 500 | 1390 | 1200 | 600 | 120 | 69.4 | 21.9 |
| 750 | 2080 | 1800 | 900 | 180 | 104 | 32.8 |
| 1000 | 2775 | 2400 | 1200 | 240 | 139 | 43.7 |
| 1500 | 4150 | 3600 | 1800 | 360 | 208 | 65.5 |
| 2000 | 5550 | 4800 | 2400 | 480 | 277 | 87.5 |
| 2500 | 6950 | 6000 | 3000 | 600 | 346 | 109 |
| 5000 | 13900 | 1200 | 6000 | 1200 | 694 | 219 |
| 7500 | 20800 | 1800 | 9000 | 1800 | 1040 | 329 |
| 10000 | 27750 | 2400 | 12000 | 2400 | 1386 | 437 |
| AMPERES = kVA x 1000/VOLTS x 1.732 | | | | | | |

Table 5.3

| Single-Phase Voltage Full-Load Current | | | | | | |
|---|---|---|---|---|---|---|
| kVA Rating | 120 | 208 | 240 | 480 | 600 | 2400 |
| 1 | 8.34 | 4.8 | 4.16 | 2.08 | 1.67 | 0.42 |
| 3 | 25 | 14.4 | 12.5 | 6.25 | 5.0 | 1.25 |
| 5 | 41.7 | 24.0 | 20.8 | 10.4 | 8.35 | 2.08 |
| 7.5 | 62.5 | 36.1 | 31.2 | 15.6 | 12.5 | 3.12 |
| 10 | 83.4 | 48 | 41.6 | 20.8 | 16.7 | 4.16 |
| 15 | 125 | 72 | 62.5 | 31.2 | 25.0 | 6.25 |
| 25 | 208 | 120 | 104 | 52 | 41.7 | 10.4 |
| 37.5 | 312 | 180 | 156 | 78 | 62.5 | 15.6 |
| 50 | 417 | 240 | 208 | 104 | 83.5 | 20.8 |
| 75 | 625 | 361 | 312 | 156 | 125 | 31.2 |
| 100 | 834 | 480 | 416 | 208 | 167 | 41.6 |
| 125 | 1042 | 600 | 520 | 260 | 208 | 52.0 |
| 167.5 | 1396 | 805 | 698 | 349 | 279 | 70.0 |
| 200 | 1666 | 960 | 833 | 416 | 333 | 83.3 |
| 250 | 2080 | 1200 | 1040 | 520 | 417 | 104 |
| 333 | 2776 | 1600 | 1388 | 694 | 555 | 139 |
| 500 | 4170 | 2400 | 2080 | 1040 | 835 | 208 |
| AMPERES = kVA x 1000/CIRCUIT VOLTAGE | | | | | | |

Table 5.4

| Radius of Conduit & Tubing Bends | | | | | |
|---|---|---|---|---|---|
| Conduit or Tubing Size | | One Shot & Full Shoe Benders | | Other Bends | |
| Metric | Trade Size | mm | inches | mm | inches |
| 16 | 1/2 | 101.6 | 4 | 101.6 | 4 |
| 21 | 3/4 | 114.3 | 4 1/2 | 127 | 5 |
| 27 | 1 | 146.05 | 5 3/4 | 152.4 | 6 |
| 35 | 1 1/4 | 184.15 | 7 1/4 | 203.2 | 8 |
| 41 | 1 1/2 | 209.55 | 8 1/4 | 254 | 10 |
| 53 | 2 | 241.3 | 9 1/2 | 304.8 | 12 |
| 63 | 2 1/2 | 266.7 | 10 1/2 | 381 | 15 |
| 78 | 3 | 330.2 | 13 | 457.2 | 18 |
| 91 | 3 1/2 | 381 | 15 | 533.4 | 21 |
| 103 | 4 | 406.4 | 16 | 609.6 | 24 |
| 129 | 5 | 609.6 | 24 | 762 | 30 |
| 155 | 6 | 762 | 30 | 914.4 | 36 |
| See NEC Table 2, Chapter 9 | | | | | |

## UNIT 6

# ELECTRICAL TERMS, COLOR CODES AND SYMBOLS

## ELECTRICAL TERMS AND DEFINITIONS

**Alternating Current:** Electric current that changes direction 60 times every second.

**Alternator:** A rotating machine that changes mechanical energy to alternating electrical current.

**American Wire Gauge:** An approved standard for sizing conducting wires.

**Ampere:** Unit of measure for current (electron flow); one ampere is an electron flow of one coulomb per second.

**Anode:** In direct current, the positive terminal such as the + post of a voltaic cell or battery.

**Apparent Power:** The product of the effective current multiplied by the effective voltage; relates to electric motors.

**Arcing:** The forming of an electric spark across the contacts of a switch or at motor or generator terminals.

**Armature:** The revolving part of a generator or electric motor.

**Armored Cable:** Electrical conductors having a flexible metal covering in addition to insulating covering around each conductor.

**AWG:** Abbreviation for American Wire Gauge; always followed by numbers indicating conductor diameter.

**Battery:** Several voltaic cells connected electrically in series; produces electrical current when placed in a circuit.

**Bond:** The continuity (unbroken path) of an electrical connection across a junction or otherwise separated conductors.

**Bonding Jumper:** A conductor that establishes electrical continuity between metal parts.

**Branch Circuit:** The part of a wiring system that extends beyond the control devices (e.g., fuse or circuit breaker).

**BX Cable:** Cable containing electrical conductors covered and protected by a flexible metal covering.

**Capacitance:** The natural property of an electrical circuit that opposes a change in voltage. It allows electrical energy to be stored in an electrostatic field.

**Capacitor Motor:** A modification of the split-phase motor; it employs a capacitor in series with its starting winding to produce a phase displacement for starting.

**Carrying Capacity:** Measure of a conductor's ability to carry current; expressed in amperes.

**Circuit Breaker:** A relay designed to open a circuit in case of an electrical overload.

**Circular Mil:** The area of a circle that is .001 inch in diameter; used to express the cross-sectional area of a conductor.

**Coil:** Conductor of electricity that has been arranged in a spiral. A magnetic field is created as current passes through the spiral.

**Commutator:** A group of bars providing connections between armature coils and brushes; found in electrical motors and generators.

**Condenser:** A device for storing electrical charges.

**Connected Load:** Electrical power that would be needed if all loads connected to the system were drawing power at the same time.

**Continuity:** Quality of a circuit that provides an unbroken path for current to travel through.

**Continuous Duty:** Operation of a circuit or electrical device under a constant load for an indefinite period.

**Controller:** A system of switches, relays, and instrumentation used to regulate voltage, current, speed, and other predetermined actions of an electrical machine or group of machines.

**Current:** Rate of electron flow through a conductor.

**Delta Connection:** Method of connecting three-phase alternators, and transformers so that the start-end of one winding is connected to the finish-end of the second winding. The circuit shape looks like the Greek letter Delta which looks like a small triangle.

**Demand Load:** Amount of power that would most likely be needed at any given time. Generally, the minimum demand load is considered to be about 35% of the connected load.

**Duplex Receptacle:** A double outlet used in house wiring circuits. It provides an easy way of connecting lamps and appliances to a circuit.

**Electrical Efficiency:** A percentage determined by comparing output power to input power.

**Electrode**: A conducting substance through which electricity travels to or from an electrical device.

**Enclosure:** Housing or box that protects persons from the energized parts of an electrical device or apparatus.

**Entrance Cap:** A weatherproof, insulated cap for terminating overhead service conductors to a building; also called a service head.

**Feeders:** One or more conductors between the service equipment and the final branch circuit overcurrent protection devices.

**Frequency:** In alternating current, the number of cycles per second (times current changes direction).

**Fuse:** An overcurrent protection device having an element that melts or destructs when subjected to high current, thus opening the circuit and stopping current before it can cause damage or over heating of the circuit.

**Generator:** Rotating machine that converts mechanical energy to electrical energy.

**Ground:** A conductor that provides connection between a circuit and the earth.

**Ground Clamp:** A fastener for connecting a ground wire or conduit to ground.

**Ground Fault:** Intentional or accidental electrical connection between a conductor and the conducting material or surfaces around it. The conducting material could be a metal housing, metal cabinet, or conduit.

**Ground Fault Circuit Interrupter:** A device placed in a circuit to detect circuit ground faults and open the circuit before it can cause damaging shock.

**Grounded**: An electrical system that has a proper connection to ground.

**Grounding Electrode:** A conductor or long rod placed in the ground for the purpose of grounding a circuit.

**Grounding Electrode Conductor:** Wire (conductor) connecting a grounding electrode to the neutral bus of the service entrance box.

**Henry:** Electrical unit of inductance.

**Hertz:** Term used to indicate frequency (cycles per second) of alternating current.

**Impedance (Z):** Resistance of alternating current circuit to both resistance and reactance. Total impedance is the vector sum of resistance and reactance.

**Inductance:** The property of an electrical circuit that opposes any change in current direction. Also, the property of a circuit that allows energy to be stored in a magnetic field (such as that created by a coil in a circuit.

**Induction Motor:** An AC motor that operates on the principle of a rotating magnetic field produced by out-of-phase currents. The rotor has no electrical connection to the energized circuit, but receives energy by magnetic transformer action from the field windings. Torque is developed as rotor current interacts with the rotating field.

**Inductor:** A coiled conductor whose purpose in an AC circuit is to oppose a change in current. Also called a choke or a coil.

**Interrupter:** A type of switch that opens and closes a circuit many times a second.

**Joule:** Unit of electric power corresponding to a watt per second.

**Kilovolt-amperes (kVA):** The product of voltage and amperage multiplied by 1000.

**Kilowatt:** One thousand watts; the same as a kilovolt-ampere.

**Kilowatt-hour:** One thousand watts/hour, a common unit of electrical energy usage.

**Kirchoff's Current Law:** At any junction of conductors in a circuit, the algebraic sum of the current is zero.

**Kirchoff's Law of Voltages:** In a simple circuit, the algebraic sum of the voltages around a circuit is equal to zero.

**Line Drop:** Voltage drop in a circuit caused by resistance.

**MHO:** Unit of conductance; the reciprocal of resistance.

**Millivolt:** One thousandth (.001) of a volt.

**Multiplier Meter:** A resistor connected to a meter in series to provide greater capacity.

**Ohm (R):** Unit of electrical resistance. The resistance offered by the passage of one ampere driven by one volt.

**Ohm's Law:** Circuit law stating that current is proportional to voltage, but inversely proportional to resistance, reactance, or impedance.

**Phase:** In the generation of alternating current, the point during rotation of a generator that electromotive force is generated.

**Polyphase:** A general term applied to an AC system having more than one phase.

**Power:** The rate at which work is done; measured in watts (W) or volt-amperes (VA).

**Power Factor (PF):** The relationship between true power and apparent power of a circuit.

**Primary:** The input windings of a transformer.

**Pull Box:** A metal box placed at sharp corners in a conduit to enable pulling of wires through the length of the conduit run.

**Reactance (I):** Opposition to alternating current caused by induction and capacitance.

**Relay:** A device located in a circuit to energize/control other parts of the circuit.

**Repulsion-Start Motor:** One that develops starting torque by the interaction of rotor currents and a single-phase stator field.

**Series Circuit:** A circuit with only a single path for current.

**Service:** The conductors that extend from the power lines to the point of entry into the building. Also called a **service drop** if overhead.

**Single-phase:** Term applied to an alternating current circuit energized by a single source of alternating electromotive force.

**Three-phase:** Term applied to an alternating current circuit energized by three alternating emf's that differ in phase by one-third of a cycle.

**Two-phase:** Term applied to an alternating current circuit energized by two alternating emf's that differ in phase by a quarter cycle.

**Volt-ampere (VA):** Product of voltage and current; the same as watts.

**Watt (W):** Unit of electrical energy.

**Watt-hour:** Electrical unit of work or electrical energy. Watt-hours equal watts times hours.

**Wye Connection:** A method of connecting three-phase alternators and transformers so that the end of each coil or winding (cont.)

has a common neutral point. The circuit configuration resembles the letter "Y".

**Zero Potential:** Having no voltage.

## LIGHTING TERMS AND DEFINITIONS

**Ballast:** Used in fluorescent and high-intensity discharge (HID) light fixtures, it gives the needed starting and operating current to the tube or bulb.

**Ballast Power Factor:** Used to express the relative efficiency in use of electric current. High-power-factor ballasts use about half the line current of normal power-factor ballasts. Thus, wiring costs, circuit loading, and certain energy costs can be reduced through use of high-power-factor ballasts.

**Foot-candle:** The amount of light from a source that reaches a surface. It is equal to the number of lumens divided by the square footage of surface the lumens are covering. In other words, a foot-candle is equal to a lumen per square foot of surface.

**Glare:** Overly bright light coming from a source or reflecting from a surface.

**HID:** Means High Intensity Discharge. This is a group of long-lasting light sources including mercury vapor, metal halide, or high-pressure sodium. All sources need a ballast to function. HIDs will operate in temperatures as low as -20° F.

**HO (high output):** A class of fluorescents similar to rapid starts. Higher level wattage up to 110 W are available for greater light output than rapid starts. Bases of HO lamps have double contacts and will only operate with HO ballasts.

**Instant Start (Slimline):** Fluorescents with ballasts that provide a high starting voltage surge. This surge quickly lights the lamp. Such lamps have a single-pin base and can be used only with instant-start ballasts.

**Lumen:** Unit of measure of amount of light given off by a lamp or other light source.

**Low-Voltage Disconnect (LVD):** Kind of circuitry used on emergency lighting fixtures. It disconnects the battery from the load when battery voltage drops below a set rating. Unnecessary deep discharge of battery is avoided, thus prolonging battery life. This does not interfere with the time rating of the fixture.

**P Class Ballast:** A ballast having internal thermal protection. It will disconnect the ballast from power source should it ever overheat. Protects against fire hazards.

**Preheat:** A class of fluorescent fixtures that require a starter. The starter heats up the lamp filaments before the ballast is allowed to supply the correct operating current through them. Preheat lamps have medium bi-pin bases. Lamps cannot be used with any other fluorescent ballast types.

**Rapid Start:** A class of fluorescents with ballasts that provide a very small amount of continuous current that keeps the filaments heated. This reduces the starting voltage surge and lights the lamps quickly. Such lamps have medium bi-pin bases. They cannot be used with any other fluorescent ballast types.

**Trigger Start:** A class of fluorescent lamps very similar to rapid starts. Found in wattages below 30 W. Rapid start or preheat lamps are used.

**UL Definitions of Damp, Dry, and Wet Locations:** Damp locations are partially protected spots such as open, roofed porches and inside locations likely to have moderate degrees of moisture (basements, barns, and cold storage enclosures). Dry locations are those not normally subject to dampness or only temporarily damp (building under construction). Wet locations are those exposed to weather or where saturation with water or other liquids is present or possible. Underground locations , concrete slabs, and masonry in contact with the earth are considered wet locations.

**Very High Output (VHO, PG, SHO):** Fluorescents much like rapid starts with wattages up to 215 W. They provide greater lumen output than rapid starts or high output units. The lamps have recessed double contact bases and will operate only with VHO, PG, or SHO ballasts.

Table 6.1

| ELECTRICAL ABBREVIATIONS | |
|---|---|
| **Abbreviation** | **The Terms** |
| A | Ampere, Armature, Anode, Ammeter |
| Ag | Silver |
| ALM | Alarm |
| AM | Ammeter |
| ARM | Armature |
| Au | Gold |
| Bk | Black |
| Bl | Blue |
| Br | Brown |
| C | Celsius |
| CAP | Capacitor |
| CB | Circuit breaker |
| CCW | Counterclockwise |
| CONT | Continuous |
| CR | Control relay |
| CT | Current transformer |
| CW | Clockwise |
| D | Diameter |
| DP | Double pole |
| DPDT | Double pole, double throw |
| EMF | Electromotive force |
| F | Fahrenheit, Field, Forward, Fast |
| FLC | Full-load current |
| FLT | Full-load torque |
| FREQ | Frequency |
| FS | Float switch |
| FTS | Foot switch |

Table 6.1 (cont)

| | |
|---|---|
| FWD | Forward |
| G | Green, Gate |
| GEN | Generator |
| GY | Gray |
| H | Transformer, primary side |
| HP | Horsepower |
| Hz | Cycles per second |
| I | Current |
| IC | Integrated circuit |
| INT | Intermediate, Interrupt |
| ITB | Inverse time breaker |
| ITCB | Instantaneous trip circuit breaker |
| K | Kilo, Cathode |
| L | Line, Load |
| LB/FT | Pounds per foot |
| LB/IN | Pounds per inch |
| LRC | Locked rotor current |
| M | Motor, Motor starter contacts |
| MED | Medium |
| N | North |
| NC | Normally closed |
| NO | Normally open |
| NTDF | Nontime delay fuse |
| O | Orange |
| OCPD | Overcurrent protection device |
| OL | Overload |
| OZ/IN | Ounces per inch |
| P | Power consumed |
| PSI | Pounds per square inch |
| PUT | Pull-up torque |

Table 6.1 (cont)

| | |
|---|---|
| R | Resistance, Radius, Red, Reverse |
| REV | Reverse |
| RPM | Revolutions per minute |
| S | Switch, Series, Slow, South |
| SCR | Silicon controlled rectifier |
| SF | Service factor |
| SP | Single-pole |
| SPDT | Single-pole, double-throw |
| SW | Switch |
| T | Terminal, Torque |
| TD | Time delay |
| TDF | Time delay fuse |
| TEMP | Temperature |
| V | Volts, Violet |
| VA | Volt-Ampere |
| VAC | Volts alternating current |
| VDC | Volts direct current |
| W | White, Watts |
| W/ | With |
| X | Transformer, secondary side |
| Y | Yellow |

Table 6.2

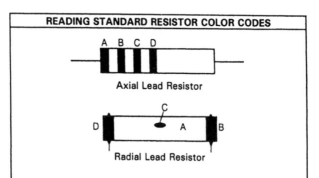

## READING STANDARD RESISTOR COLOR CODES

Axial Lead Resistor

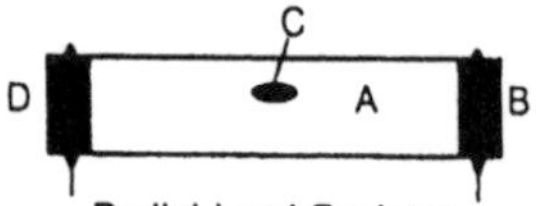


Radial Lead Resistor

| Color | A Digit 1 | B Digit 2 | C Multiplier | D Tolerance |
|---|---|---|---|---|
| Black | 0 | 0 | 1 | 1% |
| Brown | 1 | 1 | 10 | 2% |
| Red | 2 | 2 | 100 | 3% |
| Orange | 3 | 3 | 1000 | 4% |
| Yellow | 4 | 4 | 10,000 | |
| Green | 5 | 5 | 100,000 | |
| Blue | 6 | 6 | 1,000,000 | |
| Violet | 7 | 7 | 10,000,000 | |
| Gray | 8 | 8 | $10^8$ | |
| White | 9 | 9 | $10^9$ | |
| Gold | | | 0.1 (EIA) | 5% |
| Silver | | | 0.01 (EIA) | 10% |
| No color | | | | 20% |

On an axial lead resistor, a wider "A" band indicates two categories of additional information: 1) If only Band "A" is wide, then the resistor is wire wound. 2) If Band "A" is wide and there is also a blue fifth band to the right of the "D" Band, then the resistor is wire-wound and flame proof.

## COLOR CODING OF CAPACITORS

Some capacitors are color coded. The coding is basically the same as that used for resistors. One difference is that a brown color indicates a tolerance of 20% whereas tolerance of a brown-coded resistor is 1%.

Table 6.3

| STANDARD COLOR CODING AND MARKING OF CONDUCTORS FOR BRANCH CIRCUITS | |
|---|---|
| **Type of Conductor** | **Color** |
| Grounded Conductor | Continuous white or natural gray. |
| Ungrounded Conductor | Usually black or red, but can be other colors (but not white, natural gray, or green). Other colors permitted include combination of colors plus distinguishing markings. Such markings shall also be in colors other than white, natural gray, or green, and shall consist of one or more stripes of a regularly spaced series of identical markings. |
| Grounding Conductor | Continuous green plus one or more yellow stripes if not bare. |
| Heating Cables | Each unit to be marked with an identifying name or number. In addition, voltage shall be indicated by color coding as follows: (Voltages nominal.)<br>120 V--Yellow<br>208 V--Blue<br>240 V--Red<br>277 V--Brown<br>480 V--Orange |

## STANDARD SYMBOLS

Most symbols appearing on electrical plans are those approved by the American National Standards Institute (ANSI). Other symbols are sometimes used to represent special outlets. In such cases, an explanation is made on a legend that is part of a blueprint, plan or specification. Usually a small alphabet letter will accompany a special or unusual symbol.

Table 6.4

**General Outlets and Their Symbols**

| Ceiling | Wall | Type of Outlet |
|---|---|---|
| ○ | –○ | Outlet |
| Ⓑ | –Ⓑ | Blanked Outlet |
| Ⓓ | | Drop Cord |
| Ⓔ | –Ⓔ | Electrical outlet (used only when circle used alone might be mistaken for columns, plumbing symbols, etc.) |
| Ⓕ | –Ⓕ | Fan Outlet |
| Ⓙ | –Ⓙ | Junction Box |
| Ⓛ | –Ⓛ | Lamp Holder |
| Ⓛ PS | –Ⓛ PS | Lamp Holder with Pull Chain |
| Ⓢ | –Ⓢ | Pull Switch |
| Ⓥ | –Ⓥ | Outlet for Vapor Discharge Lamp |
| Ⓧ | –Ⓧ | Exit Light Outlet |
| Ⓒ | –Ⓒ | Clock Outlet |
|  | | Ceiling Outlet for Recessed Fixture |

# Electrical Lighting Plan Symbols

| | |
|---|---|
|  | 2 ft. x 4 ft. Fluorescent Lighting Fixture |
|  | 1 ft. x 4 ft. Fluorescent Lighting Fixture with 4 in. Outlet Box |
|  | Single Tube Fluorescent Strip Lighting Fixture |
|  | Two-Tube Fluorescent Strip Lighting Fixture with 4 in. Outlet Box |
|  | High Pressure Sodium Flood Light |
|  | Wall-Mounted Emergency Lighting Unit |
|  | Lighting Panel Board<br>LP = Lighting IP = Instrument<br>PP = Power HP = Heating<br>DC = Direct Current |

## Convenience Receptacles

| | |
|---|---|
|  | Duplex Receptacle |
|   | Receptacle Other Than Duplex<br>1 = Single, 3 = Triplex, etc. |
|   | Weatherproof Receptacle |
|  | 240 Volt Outlet |
|   | Range Receptacle |
|   | Switch and Convenience Receptacle |
|   | Radio and Convenience Receptacle |
|   | Ground Fault Circuit Interrupter |
|  | Special Purpose Receptacle<br>(Describe in Specs.) |
|  | Floor Receptacle |

| Switch Symbols | |
|---|---|
| $S$ | Single-Pole Switch |
| $S_2$ | Double-Pole Switch |
| $S_3$ | Three-Way Switch |
| $S_4$ | Four-Way Switch |
| $S_D$ | Automatic Door Switch |
| $S_E$ | Electrolier Switch |
| $S_K$ | Key-Operated Switch |
| $S_P$ | Switch and Pilot Lamp |
| $S_{CB}$ | Circuit Breaker Switch |
| $S_{WCB}$ | Weather-proof Circuit Breaker Switch |
| $S_{MC}$ | Momentary Contact Switch |
| $S_{RC}$ | Remote Control Switch |
| $S_{WP}$ | Weather-proof Switch |
| $S_F$ | Fused Switch |
| $S_{WF}$ | Weather-proof Fused Switch |

| Auxiliary Systems | |
|---|---|
|  | Push Button |
|  | Buzzer |
|  | Bell |
| 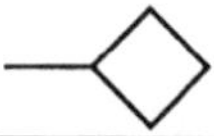 | Annunciator |
|  | Outside Telephone |
|  | Interconnecting Telephone |
|  | Telephone Switchboard |
|  | Bell-Ringing Transformer |
|  | Electric Door Opener |
|  | Fire Alarm Bell |
|  | Fire Alarm Station |

## **Auxiliary Systems** (cont.)

| | |
|---|---|
|  | City Fire Alarm Station |
| FA | Fire Alarm Central Station |
| FS | Automatic Fire Alarm Device |
| W | Watchman's Station |
| W | Watchman's Central Station |
| H | Horn |
| N | Nurse's Signal Plug |
| M | Maid's Signal Plug |
| R | Radio Outlet |
| SC | Signal Central Station |
|  | Interconnection Box |
|  | Battery |

## Electrical Power & Control Symbols

| | |
|---|---|
|  | Vertical Motor with Power & Control Combined in Single Conduit |
| 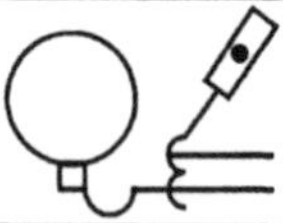 | Vertical Motor with Power & Control in Separate Conduits |
| 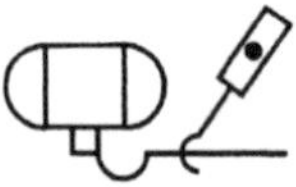 | Horizontal Motor with Power & Control Combined in Single Conduit |
| 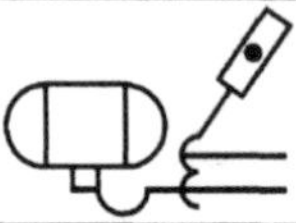 | Horizontal Motor with Power & Control in Separate Conduits |
|  | Control Station |
|  CJB | Control Junction Box |
|  | Welding Receptacle Non-Hazardous 60 A, 3 W, 4 P |
|  | Welding Receptacle - Explosion Proof 60 A, 3 W, 4 P |
|  1 | Power Receptacle 480 V Class II, 30 A |

# PLEASE SEND US YOUR FEEDBACK!

Industrial Press and Vocational Marketing Services, the co-publishers of *Harvey's Electrical Code Field Guide*, are committed to making it the very best pocket reference of its type available anywhere.

Getting your feedback is the key to making this book better and better in meeting your needs on the job or in the classroom. For example, are there any topics that might be missing or anything that might be presented more clearly?

Please mail, email, fax, or telephone your comments and suggestions to:


Industrial Press Inc.
Attn: Editorial Director
200 Madison Avenue
New York, NY 10016

Email: info@industrialpress.com
Fax: 212-545-8327
Toll-free: 888-528-7852 ext. 10


We promise to take your comments and suggestions seriously.

Thank you!